Strong, Safe, and Resilient

DIRECTIONS IN DEVELOPMENT
Environment and Sustainable Development

Strong, Safe, and Resilient

A Strategic Policy Guide for Disaster Risk Management in East Asia and the Pacific

Abhas K. Jha and Zuzana Stanton-Geddes, Editors

THE WORLD BANK
Washington, D.C.

Contents

Foreword   *xi*
Preface   *xiii*
Acknowledgments   *xv*
Editors and Contributors   *xvii*
Abbreviations   *xxv*
Key Facts about Disasters   *xxix*
Key Facts about Prevention   *xxxiii*
Note to Decision Makers   *xxxvii*

Executive Summary   1
    Where Are We Now?   1
    Where Do We Want to Be?   4
    What Needs to Be Done?   6
    How Can the World Bank Help?   8
    Notes   10
    References   10

Chapter 1   **Managing Risks in East Asia and the Pacific:**
                An Agenda for Action   13
                Key Messages for Policy Makers   13
                Where Are We Now?   13
                Where Do We Want to Be?   17
                What Needs to Be Done?   20
                How Can the World Bank Help?   24
                Notes   27
                References   28

Chapter 2   **Strengthening Institutions and Outreach to**
                Communities   31
                Key Messages for Policy Makers   31
                Where Are We Now?   31
                Where Do We Want to Be?   35
                What Needs to Be Done?   38

	How Can the World Bank Help?	40
	Notes	44
	References	44
Chapter 3	**Risk Identification**	**47**
	Key Messages for Policy Makers	47
	Where Are We Now?	47
	Where Do We Want to Be?	51
	What Needs to Be Done?	55
	How Can the World Bank Help?	58
	Notes	62
	References	62
Chapter 4	**Risk Reduction: Measures and Investments**	**65**
	Key Messages for Policy Makers	65
	Where Are We Now?	65
	Where Do We Want to Be?	67
	What Needs to Be Done?	71
	How Can the World Bank Help?	73
	References	75
Chapter 5	**Emergency Preparedness: Weather, Climate, and Hydromet Services**	**77**
	Key Messages for Policy Makers	77
	Where Are We Now?	77
	Where Do We Want to Be?	85
	What Needs to Be Done?	88
	How Can the World Bank Help?	90
	Notes	92
	References	93
Chapter 6	**Financial Protection: Risk Financing and Transfer Mechanisms**	**95**
	Key Messages for Policy Makers	95
	Where Are We Now?	95
	Where Do We Want to Be?	110
	What Needs to Be Done?	112
	How Can the World Bank Help?	112
	Notes	114
	References	115
Chapter 7	**Sustainable Recovery and Reconstruction**	**117**
	Key Messages for Policy Makers	117
	Where Are We Now?	117
	Where Do We Want to Be?	119

What Needs to Be Done?   121
How Can the World Bank Help?   122
Notes   129
References   129

Appendix A   Urbanization by Region   131

Appendix B   Large-Scale Disasters in Asia 2008–11   135

Appendix C   Vulnerability of Cities to Multiple Hazards in
              East Asia and the Pacific   137

Appendix D   Risk Identification Monitoring   139
              References   141

Appendix E   Action Plan for Building Earthquake Resilience   143
              Short Term (as Soon as Possible)   143
              Medium Term (the Next 5 Years)   143
              Long Term (5–10 Years)   144

Appendix F   Classification of Meteorological Services in East Asia
              and the Pacific   145

Appendix G   Weather and Climate Services Progress Model   147
              Observing and Forecasting Systems   147
              Weather Services Delivery   148
              Climate Services   149

Appendix H   Overview of World Bank Activities in East Asia and
              the Pacific   153
              Institutional and Capacity Building   154
              Risk Identification   154
              Risk Reduction   154
              Emergency Preparedness   155
              Financial Protection   155
              Sustainable Recovery and Reconstruction   155

Appendix I   Glossary of Key Terms   157
              Reference   160

Boxes

ES.1   Key Terms   1
1.1   Lessons from the Tohoku Earthquake   18
1.2   Approaches to Dealing with Complex Failures and Uncertainty 19
1.3   The Global Facility for Disaster Reduction and Recovery   25

1.4	Strengthening the Philippines' Resilience to Disasters	26
2.1	Impact of Cyclone Nargis in the Republic of the Union of Myanmar	32
2.2	Examples of DRM Legislation in the Region	34
2.3	Lincolnshire Mapping of Critical Assets Case Study	36
2.4	Indonesia: Using CDD Programs to Respond to Disasters	37
2.5	Partnership with ASEAN	41
2.6	Using Social Protection Mechanisms to Respond to Disasters	43
3.1	Creating Critical Infrastructure Baseline Data with Participatory Mapping	53
3.2	Integrated Flood Management Pilot in Fiji	55
3.3	Pacific Catastrophe Risk Assessment and Financing Initiative	56
3.4	What Does It Mean for Data to Be Open?	61
4.1	Building in Harm's Way	66
4.2	What Countries in East Asia and the Pacific Can Do to Prepare for the Next Big Earthquake	68
4.3	Dealing with Uncertainties: Experience from Ho Chi Minh City, Vietnam	69
4.4	Cities' Experience with a Green Infrastructure	70
4.5	Guiding Principles for Integrated Urban Flood Risk Management	73
5.1	Rainfall in Asia in 2011	79
5.2	Weather and Climate Services Progress Model	80
5.3	Case Study: Lao PDR	82
5.4	Case Study: Cambodia	83
5.5	Pacific Islands Meteorological Strategy 2012–21: Sustaining Weather and Climate Services in Pacific Island Countries and Territories	87
6.1	The Pacific Catastrophe Risk Assessment and Financing Initiative	99
6.2	National-Level Financial Catastrophe Risk Profiling	99
6.3	Toward Regional DRFI Cooperation by ASEAN Member States	105
6.4	Mongolia Index-Based Livestock Insurance Program	110
7.1	The Wenchuan Emergency Recovery Loan	123
7.2	Samoa Tsunami Post-Disaster Needs Assessment and Resilient Reconstruction	125
7.3	What Is a Post-Disaster Social Impact Analysis?	125
7.4	Safer Homes, Stronger Communities: Guiding Principles	127
7.5	Gender Concerns and the Issue of Land Titling	128
7.6	Linking Recovery with Resilient Development in Aceh	128

Figures

ES.1	East Asia and the Pacific Disasters in Economic Losses in 2011	2
ES.2	Asia's Unique Urbanization in Terms of Growth of Population, Cities, and Densities	2

ES.3   Risk Governance Capacity and World Bank Country
        Classification by Income   3
ES.4   Underinvestment of Low- and Low-to-Middle-Income
        Countries in Risk Mitigation   4
ES.5   Patterns in Jakarta between Slum and Flood-Prone Areas   5
ES.6   Making Informed Decisions to Manage Risks and Build Resilience   7
ES.7   World Bank's DRM Framework and Examples of Engagements
        in East Asia and the Pacific   9
1.1   Impact of Natural Disasters in East Asia and the Pacific in the
        Last 30 Years   14
1.2   Weather and Climate-Related Disasters and Regional Average
        Impacts, 2000–08   15
1.3   Growing Assets in Asia   16
1.4   Normalizing Losses from Nongeophysical Disasters in
        South and East Asia and Pacific Countries with
        Different Methodologies   17
1.5   Robustness to Climate Change Uncertainties   21
1.6   Formulating an Adaptive Strategy: Experience from
        the Netherlands   22
1.7   Informed Decision-Making Process to Manage Risks and Build
        Resilience   23
2.1   Post-Disaster and Pre-Disaster Spending Levels   36
3.1   Elements of Risk Identification and Risk Reduction in DRM   48
3.2   Hazard, Exposure, and Risk Maps for Papua New Guinea   49
3.3   Dynamic Decision-Making Process   51
3.4   Risk Identification across Different Levels   52
B3.1.1 Illustration of OpenStreetMap, Jakarta   53
3.5   Using Risk Assessment in Building Resilience   54
B3.3.1 Illustration of PCRAFI: Field-Surveyed Bridge in Fiji with
        Photo Validation   56
3.6   Illustration of InaSAFE Output   59
3.7   Examples of EO Information Products   60
B4.4.1 New York City-wide Costs of Combined Sewer Overflow
        Control Scenarios after 20 Years   70
6.1   Increasing Society's Financial Resilience to Disasters   96
6.2   Catastrophe Model Vendor Coverage of East Asia and the Pacific   98
6.3   Expected Loss Metrics for PICs   101
6.4   Expected Loss Metrics for ASEAN Member States   101
6.5   Disaster Losses for China, 2002–11   102
6.6   Expected Recovery and Reconstruction Liability of ASEAN
        Governments   104
6.7   Non-Life Insurance Penetration in Selected Countries in
        East Asia and the Pacific and Regionally, 2011   107
6.8   Non-Life Insurance Penetration versus Gross National
        Income (GNI) per Capita, 2011   108

6.9   Agricultural Insurance Premium, 2009   109
A.1   Growth of Cities   132
A.2   Growth of Urban Population   133
B.1   Large-Scale Disasters in Asia 2008–11   135
C.1   Cities in East Asia and the Pacific Vulnerable to Multiple
        Hazards   137
D.1   Indicator Results Related to Risk Identification HFA National
        Progress Reports   140
H.1   Map of World Bank Lending Activities in East Asia and
        the Pacific   153

Tables

4.1   Sectors Where Inertia (Lock-Ins) and Sensitivity to Climate
        Change Are Great   67
5.1   Tropical Cyclone Warning Centers in East Asia and the
        Pacific   81
6.1   Expected Emergency Losses from Disasters for PIC
        Governments   103
D.1   Select Survey Results Related to Risk Identification from
        HFA National Progress Reports   140

Foreword

We need a culture of prevention, no country can fully insulate itself from disaster risk, but every country can reduce its vulnerability. Better planning can help reduce damage—and loss of life—from disasters, and prevention can be far less costly than disaster relief and response.

—World Bank Group President Jim Yong Kim

East Asia and the Pacific (EAP) is the most disaster-stricken region in the world, with multiple challenges to building resilience.

As rapid urbanization continues, one of the main drivers of risk is poorly planned cities, which puts more people and assets in harm's way. In relative terms, the small Pacific island countries are among the most affected in the world. Average annualized losses estimated for Vanuatu are 6.6 percent of gross domestic product (GDP) and 4.3 percent for Tonga. The EAP region is also home to the second largest number of fragile and conflict-affected states after Africa, compounding the difficulties of dealing with disasters.

Disaster and disruption know no borders. In an interconnected world, even local incidents can have far reaching consequences as we saw from the large scale floods that struck Thailand in 2011. The impact of the disaster spread through industrial supply chains, and losses were felt in automobile and electric machinery production networks, regionally and globally.

In each case, disasters can wipe out decades of progress—disproportionately affecting the poor, particularly women, children, the elderly, and people with disabilities. In recent years, disaster risk management has become an increasingly important priority for the World Bank Group in its mission to end poverty. For EAP, it is critical for the region's sustained growth, as it is to continue creating opportunities for all.

Among the specific disaster risk management programs that we are pioneering or developing in the EAP region are the Indonesia Scenario Assessment for Emergencies, (InaSAFE), which is a free and publicly available tool that analyzes disaster impacts and helps keep Indonesia one step ahead on emergency planning. Participatory mapping, though OpenStreetMap tools, also allows local knowledge about critical infrastructure and social vulnerability to be included in preparedness planning. In addition, the Pacific Catastrophe Risk Assessment and Financing Initiative (PCRAFI) helps the region collect and share risk information through an open-source platform for projects on urban development, risk financing, and emergency and reconstruction planning.

In December 2011, the Philippines became the first country in the region to benefit from the World Bank's contingent financing facility—the Catastrophe Deferred Drawdown Option (Cat-DDO) in the amount of a US$500 million loan. The funding supported the World Bank's Post-Disaster Needs Assessment, with a social impact analysis that identified the needs of the most vulnerable and marginalized people hit by Typhoon Sendong.

This report captures these, and some of the new approaches and innovations being applied in the region to build a more resilient tomorrow. The report offers concrete advice to policy makers on how to reduce and manage disaster risk. It is part of an ongoing dialogue with our clients and partners, which promotes the development of innovative tools and solutions to save lives and reduce property losses in EAP.

Effective resilience requires cooperation among multiple levels of government, the private sector, civil society, and the international community. The World Bank, together with the Global Facility for Disaster Reduction and Recovery (GFDRR), works with ASEAN, JAXA, Republic of Korea NEMA, AusAID and AIFDR, SOPAC, NASA and many others to put the latest disaster risk solutions in the hands of emergency planners.

Fortunately, there is growing awareness that preventive investments in risk reduction and emergency preparedness can be extremely cost-effective and greatly reduce the impact of natural hazards.

I hope this report will help share the knowledge we have gained from our work in disaster risk management in the region, and contribute to a more resilient EAP.

Pamela Cox
Regional Vice President
East Asia and the Pacific
The World Bank Group

Preface

The fact sheets on disasters and prevention summarize some of the most recent data relevant for practitioners of disaster risk management (DRM) in the World Bank's East Asia and the Pacific region. The Note to Decision Makers introduces the key messages of this report with a focus on decision making. The executive summary provides a brief overview of the key issues, strategic goals, and recommendations for DRM in East Asia and the Pacific. Chapter 1 gives an overview of the key trends related to disaster impacts in the region. Chapter 2 focuses on cross-sectoral issues of institutional arrangements for DRM and outreach to communities. Chapters 3–7 follow the core areas of DRM: risk identification, risk reduction, emergency preparedness, financial protection, and sustainable recovery and reconstruction. The appendixes include additional information related to specific sections of the report, a glossary of key terminology, and a summary of the main activities of the World Bank East Asia and the Pacific Disaster Risk Management team.

Acknowledgments

This report outlines the challenges and opportunities as well as new priorities for the Disaster Risk Management field in East Asia and the Pacific. It takes stock of the most important activities, highlights examples of global good practice and innovative products, and makes recommendations for reducing risks and building resilience in the short, medium, and long run.

The report was produced by a team led by Abhas Jha, comprising Abigail Baca, Laura Boudreau, Henrike Brecht, Rachel Cipryk, Patricia Maria Fernandes, Olivier Mahul, Liana Zanarisoa Razafindrazay, David Rogers, Zuzana Stanton-Geddes, Zoe Trohanis, Vladimir Tsurkinov, and Eiko Wataya with inputs from Michael Bonte-Grapentin, Anna Burzykowska, Shyam KC, and Paul Procee.

The team would like to thank Bert Hofman and John Roome for their guidance; Demilour Reyes Ignacio for her support; Hitoshi Baba, Sofia Bettencourt, Chisako Fukuda, Debarati Guha-Sapir, Iwan Gunawan, Jolanta Kryspin-Watson, Jean Baptiste Migraine, Christoph Pusch, and Tomoo Ueda for their useful comments; and David Anderson for helping to edit the report.

Editors and Contributors

Editors

Abhas K. Jha is the Sector Manager for Transport, Urban, and Disaster Risk Management in East Asia and the Pacific at the World Bank. In this capacity, he is responsible for the overall technical quality control of World Bank operations, strategic staffing, and the provision of high-quality knowledge and services in these sectors to World Bank clients. Mr. Jha's core interests are smart cities, urban resilience, and the use of open data for better service delivery. He has been with the World Bank since 2001, leading the Bank's urban, housing, and disaster risk management work in Jamaica, Mexico, Peru, and Turkey as well as serving as the Regional Coordinator for Disaster Risk Management for Europe and Central Asia. Abhas has also served as Advisor to the World Bank Executive Director for Bangladesh, Bhutan, India, and Sri Lanka on issues related to urban development, infrastructure, and climate finance. Prior to joining the World Bank, Mr. Jha served for 12 years in the Indian Administrative Service of the government of India in the Federal Ministry of Finance and in the state of Bihar. Mr. Jha is the lead author of the World Bank publications *Safer Homes, Stronger Communities: A Handbook for Reconstructing after Disasters* and *Cities and Flooding: A Guide to Integrated Urban Flood Risk Management*.

Zuzana Stanton-Geddes is a Disaster Risk Management Consultant at the Sustainable Development Department in the East Asia and the Pacific Disaster Risk Management team where, since 2010, she supports operational and analytical work related to urban resilience, urban flood risk management, disaster risk financing, and gender concerns. Prior to joining the World Bank, she worked at the Friedrich Foundation in Berlin, the European Commission Permanent Representation to Germany, and IBM in Slovakia. Ms. Stanton-Geddes conducted research in Valparaiso, Chile, concerning conditions of local micro-entrepreneurs and their financing and business needs, and worked as a short-term researcher for the Center for Transatlantic Relations in Washington, DC. A native of Slovakia, Ms. Stanton-Geddes holds a master's degree in international economics and international affairs from Johns Hopkins School of Advanced International Studies (SAIS), a master's degree in European studies from Humboldt University in Berlin, and a bachelor's degree from the University of Cambridge in the UK.

Contributors

Abigail Baca is a Disaster Risk Management Specialist in the World Bank's East Asia and the Pacific Disaster Risk Management team. Since joining the World Bank's Global Facility for Disaster Reduction and Recovery (GFDRR) in 2010, she has supported multiple projects, including the Pacific Catastrophe Risk Assessment and Financing Initiative, Building Urban Resilience in East Asia, and GFDRR's Open Data for Resilience Initiative (OpenDRI), including the InaSAFE development in partnership with the Australia-Indonesia Facility for Disaster Reduction (AIFDR). Prior to that she worked as a natural catastrophe risk modeler and product manager, gaining six years of international, multiperil modeling experience at Risk Management Solutions Inc. (RMS). Ms. Baca served as a vulnerability engineer for multiple projects, including probabilistic earthquake and climate risk models for the Americas and Europe. She earned a BS in civil and environmental engineering from Stanford University and an MS in structural engineering from the University of California, San Diego.

Michael Bonte-Grapentin is a Disaster Risk Management and Climate Change Adaptation Specialist with the World Bank Pacific Department. He brings with him more than 12 years of experience in disaster risk management. Before joining the Bank, he was leading the Risk Reduction Team at SOPAC, the Pacific regional organization mandated with building disaster risk management capacity. His work focused on strengthening technical institutions in the Pacific in providing flood and multihazard warning services by conducting risk and technical post-disaster assessments. Mr. Bonte-Grapentin brings with him private sector experience, working previously for German Telekom on net-information and operation support systems. He was also a researcher on alpine hazards for the Free University of Berlin. He holds a master's degree in geography with an international award-winning thesis on debris avalanche initiation.

Laura Boudreau is a Disaster Risk Financing and Insurance Analyst working for the World Bank's Global Facility for Disaster Reduction and Recovery (GFDRR) Disaster Risk Financing and Insurance (DRFI) Program, where she has been working since 2010. Ms. Boudreau provides technical assistance on disaster risk financing and insurance in low- and middle-income countries, specifically technical assistance for sovereign disaster risk financing and insurance for middle-income countries, with a regional focus on Latin America and the Caribbean, although she is also involved in the DRFI Program's larger portfolio. Ms. Boudreau has formerly worked as a research assistant at the Wharton Risk Management and Decision Processes Center and for the Europa Reinsurance Facility Ltd. She has a background in applied economics, risk analysis, and globalization, and she graduated with a business degree from the Wharton School of the University of Pennsylvania.

Henrike Brecht is a Disaster Risk Management Specialist in the East Asia and Pacific Region of the World Bank. She leads the World Bank's disaster risk management agenda in a number of Southeast Asian countries and has conducted in-depth research in the areas of disaster risk assessments and flood risk management. She has published articles in journals such as the *Journal of Environment and Development* and *Social Science Computer Review*. Current research interests include water resource management and hydro-meteorological services. In the past, Ms. Brecht worked in several different settings, including the UN, research institutions, and the private sector. Her responsibilities ranged from overseeing the program portfolio at the Global Facility for Disaster Reduction and Recovery (GFDRR) and working on recovery planning at the Louisiana Hurricane Center after Hurricane Katrina in New Orleans to researching synergies between watersheds and settlement planning at the United Nations High Commissioner for Refugees (UNHCR) in Kenya, Sudan, and Switzerland. Ms. Brecht holds an MS degree in environmental science and a PhD in disaster risk management.

Anna Burzykowska is an Earth Observation Projects Specialist who joined the World Bank in 2011 as a secondee from the European Space Agency (ESA). She is involved in the initiative called Earth Observation for Development, dedicated to introducing innovative satellite services to various areas of the World Bank's Sustainable Development Network (Urban Development, Disaster Risk Management, Water Resources Management, Forestry, Agriculture, Coastal Zones Management, and Climate Information Services). Ms. Burzykowska held positions at the ESA Washington Office and the ESA Directorate of Earth Observation Programmes in Frascati, Italy, beginning in 2008. Before joining ESA, she had been working for the Polish Special Representative to the UNFCCC COP14 as an expert on matters related to the Subsidiary Body for Scientific and Technological Advice and at the Space Research Center in Warsaw. Ms. Burzykowska holds degrees from Warsaw University, University of Leiden, and King's College London and was a visiting researcher at George Washington University's Center for Science and Technology Policy.

Rachel Cipryk is Disaster Risk Management Analyst on the World Bank's East Asia and the Pacific Disaster Risk Management team. She specializes in risk management from a social perspective and has pursued this through working on disaster risk management, social protection, livelihoods, and food security programs. A primary focus of her operational and analytical work is in linking social protection with disaster risk management and climate change programming. Ms. Cipryk's operational experience includes Ethiopia's Productive Safety Net Programme Donor Coordination Team, as a consultant engaging in research and program design; the Famine Early Warning Systems Network (FEWS NET), developing livelihoods knowledge products for food security and early warning analysis; and projects in the Pacific with the Disaster Risk Management team at the World Bank. She earned a master's degree in poverty and development from the Institute of Development Studies at IDS in the UK and an undergraduate degree in political science.

Patricia Fernandes is a Social Development Specialist in the East Asia and Pacific Region of the World Bank. She joined the World Bank in 2008, working on Community-Driven Development operations in the Philippines and participatory approaches to rural poverty reduction in China. She has experience in leading qualitative research and in social analysis of natural disasters. Prior to joining the World Bank, Ms. Fernandes worked on development programs in Angola and Mozambique with UNICEF and in Kosovo with the UN.

Shyam KC is a Disaster Risk Management Specialist working in the East Asia and Pacific Region at the World Bank. He started his career in development as a civil engineer with the government of Nepal and then moved on to the Swiss Association for International Cooperation, where he managed a community-based integrated water resources program in Nepal, focusing on water supply and sanitation. He returned to graduate school in 2001, where as a sociology student he studied the intricate relationship between society and infrastructure for his doctorate, and society and conflict for his master's degree. Prior to joining the East Asia and the Pacific Disaster Risk Management team in 2010, Mr. KC worked in the World Bank's Global Facility for Disaster Reduction and Recovery (GFDRR) as a Disaster Risk Management Specialist.

Olivier Mahul is the Program Coordinator of the World Bank's Disaster Risk Financing and Insurance Program (DRFI), which aims to mainstream disaster risk financing within the disaster risk management and climate change adaptation agenda. Mr. Mahul also coordinates the Insurance for the Poor Program, which provides client countries with technical assistance on the design and implementation of insurance products. Since joining the World Bank in 2003, Mr. Mahul has been involved in designing and implementing disaster risk financing solutions in several developing countries, including Colombia, Costa Rica, India, Indonesia, Mongolia, and Vietnam. Mr. Mahul is one of the architects of the Caribbean Catastrophe Risk Insurance Facility, and he is currently leading a similar initiative in the Pacific region. Mr. Mahul holds a PhD in economics of risk and insurance from the Toulouse School of Economics and a post-doctorate from Wharton Business School and the University of California at Berkeley. Mr. Mahul has authored more than 30 publications in international journals and won several academic awards. He recently co-authored the books *Catastrophe Risk Financing in Developing Countries: Principles for Public Intervention*, with J. David Cummins, and *Government Support to Agricultural Insurance: Challenges and Options for Developing Countries*, with Charles Stutley.

Paul Procee works as the Lead Urban Specialist at the World Bank office in China. He is an environmental engineer with more than 15 years of experience in urban environmental and infrastructure management. He joined the World

Bank in 1999 and has been working on a wide range of projects related to urban infrastructure, water supply, sanitation, solid waste, transport, and disaster risk management. He started working in the World Bank Institute and moved to the Latin American Region in 2006. In 2010 Mr. Procee joined the Urban Rural Integration team in the China Office in Beijing, where he manages a variety of urban infrastructure projects and coordinates the disaster risk management activities of the Bank in China. Among others, he prepared a policy note on integrated flood risk management for the Chinese government, and he currently manages the Wenchuan Earthquake Recovery Project and a number of urban infrastructure and flood control projects. He is also preparing a publication on climate risk management and adaptation.

Liana Zanarisoa Razafindrazay is a Consultant on the East Asia and the Pacific Disaster Risk Management team. She specializes in open data and geographic information systems (GISs) applied to disaster risk management, bringing with her five years of experience in the field. Prior to joining the Bank, she was a senior staff associate at the Center for International Earth Science Information Network (CIESIN) at the Earth Institute of Columbia University, where she was part of a multidisciplinary science and geospatial specialist team. Ms. Razafindrazay provided support on various international projects, undertaking geospatial analysis and data visualization in disaster risk assessment, project monitoring and evaluation, socioeconomic data production, climate change, and migration. She also provided technical support for the UN Office for the Coordination of Humanitarian Affairs (OCHA) at the Reliefweb Map Center Unit, managing information related to ongoing disasters. A native of Madagascar, Ms. Razafindrazay holds master's degrees in both geography and sustainable development from Pantheon-Sorbonne University and CERDI in Clermont-Ferrand, France.

David Rogers is the President of the Health and Climate Foundation (HCF), an international nonprofit organization dedicated to finding solutions to climate-related health problems and supporting partnerships between health and climate practitioners. Prior to founding HCF, Dr. Rogers held various appointments in government, the private sector, and academia. These include Chief Executive, UK Met Office; Vice President, Science Applications International Corporation; Director, Office of Weather and Air Quality, U.S. National Oceanic and Atmospheric Administration; Director, Physical Oceanography, Scripps Institution of Oceanography; and Associate Director, California Space Institute, University of California at San Diego. Currently, he is a consultant to the World Bank on modernizing national meteorological and hydrological services. Dr. Rogers has a PhD (1983) from the University of Southampton and a bachelor of science degree (1980) from the University of East Anglia, UK. He has published extensively in the fields of oceanography, meteorology, climate, environment, and organizational development.

Zoe Trohanis works in the Latin America and the Caribbean region of the World Bank as a Senior Urban Specialist, working on urban and regional development operations in Bolivia and Peru. Prior to this position, she worked in the East Asia and Pacific (EAP) region as a Senior Infrastructure Specialist, with more than 10 years of experience at the Bank in urban and disaster risk management. Since joining EAP in 2008, she has worked on urban development projects and done technical assistance and analytical work in Indonesia, the Philippines, and Vietnam, with a focus on slum upgrading, urban poverty and gender concerns, and urban resilience. Examples include the Indonesia National Community Empowerment Program in Urban Areas, the Metro Manila Flood Masterplan, and the Vietnam Urban Upgrading Project and National Urban Upgrading Strategy. She has also been leading the Bank's dialogue on disaster risk management with the government of the Philippines as task team leader of the Disaster Risk Management Development Policy Loan with a Catastrophe Deferred Drawdown Option (Cat-DDO) and has worked on post-disaster recovery programs in China, Indonesia, Pakistan, and Sri Lanka. Prior to her position in EAP, Ms. Trohanis was a World Bank Urban Development anchor. She has a BA from the University of Wisconsin Madison and an MA from American University's School of International Service.

Vladimir Tsirkunov works as a Senior Environmental Engineer in the World Bank's Global Facility for Disaster Reduction and Recovery (GFDRR). Mr. Tsirkunov has more than 30 years of scientific, applied technical, and project management experience in environmental and natural resource management. Prior to joining the World Bank in 1994, he was a Head of the Laboratory of the Supervision of the USSR (Russian) System of Hydrochemical Monitoring and Water Quality Data Collection. For the past 18 years he has held progressively more responsible positions with the World Bank related to the preparation, appraisal, and operational supervision of environmental investments, technical assistance, and Global Environment Facility (GEF) projects. Beginning in 2004 he helped prepare new investment operations and analytical products supporting the improvement of weather, climate, and hydrological services, initially in Europe and Central Asia and later in East Asia, Africa, South Asia, and other regions. Since 2011, Mr. Tsirkunov has been leading the GFDRR's Program of Strengthening Hydrometeorological Agencies and Decision Support Systems, which functions as a service center providing analytical, advisory, and implementation support for World Bank teams.

Eiko Wataya is the Knowledge Management Officer for the East Asia and the Pacific Disaster Risk Management team. She has more than 8 years of experience in distance/conventional type learning, knowledge-sharing program development, and implementation of institutional capacity-building. Ms. Wataya joined the World Bank in 2004 and has been working on a range of programs, including in the areas of disaster risk management, SMEs, and productivity improvement. She started working in the Tokyo Development Learning Center of the World Bank as Program Coordinator and moved to the East Asia and the Pacific Disaster Risk Management team in 2010 to develop a knowledge management strategy and lead planning and delivery of regional disaster risk management communities of practice activities in collaboration with key partners inside and outside the World Bank. Prior to joining the World Bank, Ms. Wataya worked at the Japan Bank for International Cooperation as a Country Officer for Indonesia and Malaysia on lending projects, technical assistance, and analytical work in the urban sector.

Abbreviations

AADMER	ASEAN Agreement on Disaster Management and Emergency Response
AAL	average annualized loss
ADB	Asian Development Bank
ADPC	Asian Disaster Preparedness Center
ADRC	Asian Disaster Reduction Center
AIFDR	Australia-Indonesia Facility for Disaster Reduction
ARPDM	ASEAN Regional Programme on Disaster Management
ASEAN	Association of Southeast Asian Nations
AusAID	Australian Agency for International Development
BNPB	Indonesian National Disaster Management Agency
BPBD	Province of Jakarta Disaster Management Agency
Cat-DDO	Catastrophe Deferred Drawdown Option
CCA	climate change adaptation
CDD	community-driven development
CMA	China Meteorological Administration
DHRW	Department of Hydrology and River Works
DMH	Department of Meteorology and Hydrology
DOM	Department of Meteorology
DRFI	disaster risk financing and insurance
DRM	disaster risk management
DRR	disaster risk reduction
DRRM	disaster risk reduction and management
EAIG	East Asia Infrastructure Growth Fund
ECMWF	European Centre for Medium-Range Weather Forecasts
EO	earth observation
EPS	ensemble prediction system
ERL	emergency recovery loan
ESA	European Space Agency
EWS	early warning system
FID	Madagascar Social Development Fund
GDP	gross domestic product
GFDRR	Global Facility for Disaster Reduction and Recovery

GIS	geographic information system
GTS	Global Telecommunication System
HFA	Hyogo Framework for Action
HKO	Hong Kong Observatory
Hydromet	hydrometeorological
IBLIP	Index-Based Livestock Insurance Program
IBRD	International Bank for Reconstruction and Development
IDA	International Development Association
InaSAFE	Indonesia Scenario Assessment for Emergencies
InSAR-based PSI	interferometric synthetic aperture radar–based persistent scatterer interferometry
IPCC	Intergovernmental Panel on Climate Change
IRR	internal rate of return
ISDR	International Strategy for Disaster Reduction
JMA	Japan Meteorological Agency
JRF	Java Reconstruction Fund
KMA	Republic of Korea Meteorological Administration
MHEWS	Multi-Hazard Early Warning System
MoC	Memorandum of Cooperation
MRC	Mekong River Commission
NAB	National Advisory Board
NGO	nongovernmental organization
NHS	National Hydrometeorological Services
NMS	National Meteorological Service
NWP	numerical weather prediction
OCHA	Office for the Coordination of Humanitarian Affairs
OpenDRI	Open Data for Resilience Initiative
PCRAFI	Pacific Catastrophe Risk Assessment and Financing Initiative
PDNA	post-disaster needs assessment
PHRD	Japan Policy and Human Resources Development
PIC	Pacific island country
PMC	Pacific Meteorological Council
PNPM Mandiri	Indonesian National Program for Community Empowerment Mandiri
PPP	public-private partnership
PSDR	Rural Development Support Project
RSMC	Regional Specialized Meteorological Center
SAR	Special Administrative Region
SIA	social impact assessment
SOPAC	Secretariat of the Pacific Community
SWFDP	Severe Weather Forecasting Demonstration Project
TCWC	Tropical Cyclone Warning Center
UNEP	United Nations Environment Programme

UNISDR	United Nations International Strategy for Disaster Reduction, United Nations Office for Disaster Risk Reduction
WERP	Wenchuan Earthquake Project
WMO	World Meteorological Organization

Key Facts about Disasters

East Asia and the Pacific is the most disaster-stricken region in the world, suffering from small recurrent as well as rare high-impact events.

- In 2011, disaster losses amounted to **US$380 billion.** East Asia sustained 80 percent of these losses in the first nine months.[1]
- More than **1.6 billion** people have been affected by disaster in East Asia and the Pacific since 2000 (EM-DAT 2012).[2]
- The vulnerability to flooding will increase in Asia, with a projected **410 million urban** Asians at risk of coastal flooding by 2025 (ADB 2012). East Asia and the Pacific accounts for about 40 percent of the total number of floods worldwide over the past 30 years.

Disasters disproportionally affect the poor, vulnerable, and marginalized—including women, children, the elderly, and people with disabilities.

- **Women are more likely than men to die from natural disasters** when their socioeconomic status is low (Neumayer and Plümper 2007). Women represented an estimated **70 percent** of casualties after the 2004 Indian Ocean tsunami in Banda Aceh.
- **Disasters can push affected households further into debt** with the poor carrying the greatest debt burden. Two years after Cyclone Nargis in the Republic of the Union of Myanmar, average maximum debt across villages increased. Average maximum debt of laborers and fishermen more than doubled, and that of small farmers was almost twice as high.

East Asia and the Pacific follows a unique urbanization pattern in terms of growth of population, cities, and densities.

- From 1980 to 2010, Asia added over **one billion** people to its cities—more than all other regions combined—and another billion inhabitants will live in cities by 2040 (UN 2011a).
- Urban Asia has high population densities and most of the world's megacities. In 2010 Asia was home to over half, or 12 of 23, of the world's megacities; by

2025 the number of megacities in Asia is expected to increase to **21 of a global total of 37** (ADB 2012; UN 2011b).
- Globally, **informal settlements are growing at a much faster pace than cities themselves** (Perlman 2010). In absolute terms Asia is home to more than half the world's slum population (UN Habitat 2006). In metropolitan Manila, for example, it is estimated that about **800,000 people,** mostly informal settlers, live in high-risk to very high-risk areas.

East Asia and the Pacific is exposed to large fiscal impacts on public expenditure.

- Cambodia, the Lao People's Democratic Republic, the Republic of the Union of Myanmar, the Philippines, and Vietnam face particularly high annual average expected losses relative to the size of their economies. Cambodia, Lao PDR, and the Philippines could face costs totaling **18 percent or more of total public expenditure** in the event of a 200-year disaster.

Small Pacific islands are particularly vulnerable to the impacts of disasters.

- In the Solomon Islands, the 8.1 magnitude earthquake followed by a tsunami that hit in April 2007 caused losses estimated **at 95 percent of the government's budget** and created a short-term liquidity crunch until donor assistance arrived. The tsunami that hit Samoa in September 2009 caused losses estimated at **22 percent of national gross domestic product (GDP).**
- The average annual losses from tropical cyclones and earthquakes are estimated to be as high as **6.6 percent** of national GDP in countries such as Vanuatu.

Notes

1. Figures from Munich Re (2012). Note that estimates differ: EM-DAT (2012) estimates damage losses at US$366.1 billion for natural disasters that occurred in 2011. The Swiss Re estimate is US$370 billion.
2. Calculation based on EM-DAT data accessed online in September 2012.

References

ADB (Asian Development Bank). 2012. "Green Urbanization in Asia Key." Special chapter in *Indicators for Asia and the Pacific 2012.* Manila: Asian Development Bank.

EM-DAT. 2012. The International Disaster Database. Centre for Research on the Epidemiology of Disasters (CRED), Brussels. Accessed September 2012. http://www.emdat.be/database.

Munich Re. 2012. *NatCatSERVICE Natural Catastrophes Worldwide 2011.* Percentage distribution (online brief) Münchener Rückversicherungs-Gesellschaft, Geo Risks Research. http://www.munichre.com/app_pages/www/@res/pdf/NatCatService/

annual_statistics/2011/2011_mrnatcatservice_natural_disasters2011_perc_distrib_ event_by_type_en.pdf.

Neumayer, E., and T. Plümper. 2007. "The Gendered Nature of Natural Disasters: The Impact of Catastrophic Events on the Gender Gap in Life Expectancy, 1981–2002." *Annals of the Association of American Geographers* 97 (3): 551–66. http://www2.lse .ac.uk/geographyAndEnvironment/whosWho/profiles/neumayer/pdf/ Disastersarticle.pdf.

Perlman, J. 2010. *Favela: Four Decades of Living on the Edge of Rio.* Oxford: Oxford University Press.

UN (United Nations). 2011a. *United Nations Global Assessment Report on Disaster Risk Reduction.* New York: United Nations.

———. 2011b. *World Population Prospects: The 2010 Revision.* Population Division of the Department of Economic and Social Affairs of the United Nations Secretariat, United Nations, New York. http://esa.un.org/wpp/index.htm.

UN (United Nations) Habitat. 2006. "Slum Trends in Asia." Online brief. United Nations, New York. Accessed September 2012. http://www.unhabitat.org/documents/media_ centre/APMC/Slum%20trends%20in%20Asia.pdf.

Key Facts about Prevention

Hazard mitigation is most effective when based on inclusive, long-term planning developed before a disaster strikes (U.S. Department of Homeland Security 2012).

- Cost-benefit analyses found that **every US$1 spent on mitigation** saved countries **US$3–US$4.1**.[1]

Investing in disaster preparedness and the right balance of structural and nonstructural measures can be highly cost-effective.

- During a flood protection project in Argentina, an internal rate of return (IRR) was estimated at **12–79 percent** for flood mitigation measures. The overall project IRR falls from **20.4 to 7.5 percent** if the project start date is initiated five years later, making a case for not delaying the project.
- Following Hurricane David in Dominica in 1979, **4.2 percent** of the total construction cost was spent on seaport reconstruction, whereas only an additional **1.9 percent** would have been sufficient to mitigate the losses incurred after the disaster.

Strengthening hazard forecast and hydrometeorological services is a no-regret investment with a high cost-benefit ratio.

- In China the cost-benefit ratio of strengthening national meteorological services can range between **1:35** and **1:40**.

In areas of risk, it is more cost-effective to strengthen existing school buildings than to entirely rebuild them.

- Experience from a seismic preparedness project in Istanbul, Turkey, shows that **five to seven schools** can be strengthened for the cost of **one new building**. In Colombia, probable maximum loss for a 1-in-1,000-year earthquake for a retrofitted school would be **4 percent** of the asset value compared to **30 percent** without retrofits.
- In Jamaica, had hurricane-resistant features been integrated in building design, the cost would have been **1 percent** of the total building costs compared to a benefit of **35–40 percent** for mitigating impacts of a hurricane of a similar magnitude to Hurricane Gilbert in 1988 (Pereira 1995).

Coral reefs and mangroves are among the most valuable ecosystems and provide many benefits.

- Annual values per square kilometer for coral reefs and mangroves were calculated at US$100,000–US$600,000 for reefs and US$200,000–US$900,000 for mangroves (Wells et al. 2006).
- In Vietnam planning and protection of 12,000 hectares of mangroves cost US$1.1 million, with a benefit of reduction of dike maintenance cost **by US$7.3 million per year.** Deaths were eliminated, and livelihoods of 7,750 families were positively affected (Kay and Wilderspin 2002).
- Preventing damage to coastal infrastructure and flooding, mangroves reduce wave height by as much as **66 percent over 100 meters of forest** (McIvor et al. 2012).

Restoring natural ecosystems can be more cost-effective than engineered solutions.

- In the Philippines, the benefit-cost ratio for protecting the city of Angeles from lahar flows was calculated as **30:1** for rain forestation farming, **14.7:1** for bamboo plantations, and **3.5:1** for river channel improvement (Dedeurwaerdere 1998).
- Every **2.7 miles** of marshland that a hurricane has to travel reduces the storm surge by **one foot.** If Katrina had struck in 1945 instead of 2005, the surge that reached New Orleans would have been as much as **5–10 feet shallower** (Tidwell 2005).

Using existing social protection and community-driven development (CDD) interventions can substantially reduce disaster response costs.

- CDD approaches for smaller-scale disaster preparedness investments have proven to be consistently cost-effective. In the Philippines, cost savings ranged from **8 percent for school buildings** to **76 percent for water supply investments,** when compared with traditionally implemented infrastructure (Araral, and Holmemo 2007).

Open risk information and data enable stakeholders to make better decisions.

- Opening up government data in the United Kingdom will create an estimated **£6 billion** in additional value for the U.K. economy from businesses that will create added value services using the information (Pollock 2009).
- In the United States, mandatory public disclosure of higher flood risk areas in North Carolina resulted in a differential of a **7.3 percent** decrease in property values and a corresponding increase in insurance premium prices (Bin et al. 2006; Lall and Deichmann 2009).

Note

1. U.S. Department of Homeland Security (2012) cites Congress of the United States, Congressional Budget Office, Potential Cost Savings from the Pre-Disaster Mitigation Program (Washington, DC, 2007); and National Institute of Building Sciences, The

Multihazard Mitigation Council, *Natural Hazard Mitigation Saves: An Independent Study to Assess the Future Savings from Mitigation Activities* (Washington, DC, 2005).

References

Araral, E., and C. Holmemo. 2007. "Measuring the Costs and Benefits of CDD: The KALAHI-CIDSS Project Philippines." World Bank Social Development Papers 102, World Bank, Washington, DC. http://siteresources.worldbank.org/EXTSOCIALDEVELOPMENT/Resources/244362-1164107274725/3182370-1164201144397/3187094-1173195121091/SDP-102-Jan-2007.pdf.

Bin, O., J. B. Kruse, and C. E. Landry. 2006. *Flood Hazards, Insurance Rates, and Amenities: Evidence from the Coastal Housing Market*. Department of Economics, East Carolina University, Greenville, NC. http://www.ecu.edu/cs-educ/econ/upload/ecu0603.pdf.

Dedeurwaerdere, A. 1998. "Cost-Benefit Analysis for Natural Disaster Management: A Case-Study in the Philippines." Centre for Research on the Epidemiology of Disasters Working Paper 143, Université Catholique de Louvain, Louvain-La-Neuve, Belgium.

Kay, R., and I. Wilderspin. 2002. "Box 4.4: Mangrove Planting Saves Lives and Money in Vietnam." In *World Disaster Report Focus on Reducing Risk*, 95. Geneva: International Federation of Red Cross and Red Crescent Societies (IFRCRCS). http://www.ifrc.org/Global/Publications/disasters/WDR/32600-WDR2002.pdf.

Lall, S. V., and U. Deichmann. 2009. "Density and Disasters: Economics of Urban Hazard Risk." *World Bank Research Observer* 27 (1): 74–105.

McIvor, A. L., I. Möller, T. Spencer, and M. Spalding. 2012. "Reduction of Wind and Swell Waves by Mangroves." Natural Coastal Protection Series: Report 1. Cambridge Coastal Research Unit Working Paper 40. The Nature Conservancy and Wetlands International. http://www.wetlands.org/WatchRead/Currentpublications/tabid/56/mod/1570/articleType/ArticleView/articleId/3353/Default.aspx.

Pereira, J. 1995. "Costs and Benefits of Disaster Mitigation in the Construction Industry." Paper presented at the Caribbean Disaster Mitigation Project Workshop, Trinidad. March 14–16. http://www.oas.org/cdmp/document/papers/pereira.htm.

Pollock, R. 2009. "The Economics of Public Sector Information." Cambridge Working Papers in Economics 0920, Faculty of Economics, University of Cambridge, Cambridge, U.K. http://www.rufuspollock.org/economics/.

Rohan, K., and I. Wilderspin. 2002. "Box 4.4: Mangrove Planting Saves Lives and Money in Vietnam." In *World Disaster Report Focus on Reducing Risk*, 95. Geneva: International Federation of Red Cross and Red Crescent Societies (IFRCRCS). http://www.ifrc.org/Global/Publications/disasters/WDR/32600-WDR2002.pdf.

Tidwell, M. 2005. "Goodbye, New Orleans: It's Time We Stopped Pretending." AlerNet online article December 5, 2005. Accessed September 2012. http://www.alternet.org/story/29274/goodbye,_new_orleans?page=2%2C1&paging=off.

U.S. DHS (United States Department of Homeland Security). 2012. *Survey of Hazard Mitigation Planning*. Washington, DC. http://www.oig.dhs.gov/assets/Mgmt/2012/OIG_12-109_Aug12.pdf.

Wells, S., C. Ravilious, and E. Corcoran. 2006. *In the Front Line: Shoreline Protection and Other Ecosystem Services from Mangroves and Coral Reefs*. Cambridge, U.K.: UNEPA World Conservation Monitoring Centre.

Note to Decision Makers

Why Does Disaster Risk Management Matter?

Disaster risk management (DRM) is essential in the fight against poverty. Disasters can, in an instant, wipe out decades of hard-fought poverty reduction and development gains and push countless households into poverty. Disasters disproportionally affect the poor: vulnerable and marginalized groups, including women, children, the elderly, and people with disabilities, are at particular risk.

East Asia is rapidly urbanizing, and cities are becoming disaster hotspots. Unplanned or poorly planned urbanization that puts more people and assets in harm's way is the single largest driver of disaster risk. From 1980 to 2010, Asia added over one billion people to its cities—more than all other regions combined—and another billion inhabitants are expected to live in urban areas by 2040.[1] Much of this growth will take the form of informal settlements located in areas at risk given the limited availability and affordability of land in these cities, placing a significant number of particularly vulnerable households at risk. At the same time, many East Asian cities are part of complex global supply chains where single-event failures can lead to cascading disasters reaching beyond the boundaries of an urban area, country, or region.

Preventive investments in risk reduction and emergency preparedness can be extremely cost-effective and can greatly reduce the impact of natural hazards. Public investments, such as early-warning systems, retrofitting of critical infrastructure at risk, and mainstreaming systematic risk assessments into relevant public investment planning processes, can have a positive effect on countries' efforts to reduce poverty and promote sustainable economic growth.

Why Is Mainstreaming DRM into Development Difficult?

Inadequate institutional arrangements and poor coordination across agencies and levels of government as well as the private sector and civil society stall the process of mainstreaming DRM into development. Institutions and stakeholders often lack incentives to cooperate and invest in preparedness.

There is deep uncertainty about future disaster and climate risks, challenging our ability to adapt to new developments and changing the physical and natural environment. Scarcity of resources and often stark trade-offs among competing priorities can lead to wishful thinking ("Not in my term of office") and postponing difficult choices to the future.

What Needs to Be Done?

Decision makers can make a significant difference by effectively managing disaster risk and building resilience. With education and communication, preparedness, and investments, urbanization can be channeled as a tremendous positive force for development. Better urban planning, coordination, and development provide a unique opportunity to make a lasting positive impact on the lives of many people and specifically address the needs of urban poor who face some of the highest risks.

Hazards are natural, disasters are manmade. Every natural hazard does not necessarily need to turn into a disaster. By decreasing disaster exposure and vulnerability through systematic assessments and communication of risks, better land-use planning, and many other practical measures, the impacts of natural hazards can be reduced significantly.

We cannot build our way to safety. It is necessary to recognize that the risks of disasters cannot be entirely eliminated. Residual risks and uncertainties need to be managed by investing in the right balance of structural and nonstructural measures. Nonstructural measures, such as early warning systems, can be highly cost-effective. At the same time, we need to plan for failure by considering different scenarios, especially within complex systems and networks.

Governments should prioritize actions based on informed decisions about the level of risk to reduce the risks from disasters. Informed decision making includes an assessment of the levels of risk, determines whether it should be reduced, transferred, or managed, and decides the best ways to do so given the existing capacity and available instruments. The level of risk, a cost-benefit analysis of possible interventions, existing capacity, and affordability, should guide the decision makers in the prioritization of the recommendations presented in this report, which form part of a phased, incremental, and iterative strategy to DRM.

In the short term, investments that have a high positive cost-benefit ratio with immediate and significant payoffs include strengthening emergency preparedness and early warning systems, as well as improving institutional arrangements and capacities by working closely with communities and local-level institutions. Investing in reliable risk information at the national, regional, city, and community levels helps to assess and communicate the socioeconomic and fiscal impacts of disasters and formulate effective DRM strategies.

In the medium to long term, one of the biggest challenges is getting the balance right between structural and nonstructural investments, as well as within structural investments and between "gray" concrete investments and cost-effective "green" infrastructure, such as building mangroves, wetland buffers, and coastal restoration. "Getting the balance right" includes a stronger focus on social protection and community-driven development programs and ecosystem management, as well as developing a comprehensive disaster risk-financing strategy with ex ante and ex post instruments. Sharing and effectively communicating risk information between agencies across different levels and sectors and the public and the private sectors enables stakeholders to make better-informed decisions and strengthens the collective resilience of a community.

In the long term, effective enforcement of risk-based land-use planning and building codes in building public infrastructure in safe areas, together with providing necessary services and coping mechanisms to benefit the most vulnerable and poor, should be a priority. To deal with the risks and deep uncertainties linked to natural hazards and climate change, we need to focus on robust low-regret solutions that can bring benefits under a range of scenarios. This underlines the need to promote DRM as an iterative process, in which lessons learned and new technologies can help us to better adapt to changed circumstances. Promoting the development of risk insurance markets through public-private partnerships and the development of enabling regulatory infrastructure can assist the development of a cost-effective, affordable, and sustainable insurance market and facilitate disaster risk pooling with a more diversified portfolio.

How Can the World Bank Help?

The World Bank supports countries around the world in mainstreaming a comprehensive and integrated approach to DRM into development. With its overarching mission to fight poverty, the World Bank's DRM efforts aim to build resilient communities. Between fiscal years 2006 and 2012, the International Bank for Reconstruction and Development (IBRD) and the International Development Association (IDA) committed an estimated US$11.7 billion to 181 projects related to DRM (GFDRR 2012).[2] Before 2006 the largest disaster projects always focused on reconstruction. Today the Bank finances large disaster preparedness projects, for example, the Disaster Risk Management Catastrophe Drawdown Option Loan for the Philippines (US$500 million), the India National Cyclone Risk Mitigation Project (US$255 million), the Jakarta Urgent Flood Mitigation Project (US$131 million), and the Istanbul Seismic Risk Mitigation and Emergency Preparedness Project (US$150 million).

In East Asia and the Pacific, the World Bank supports a range of low- to upper-high-income countries in developing effective disaster preparedness and response measures. Paying close attention to countries' singular context, the World Bank provides analytical and advisory services, helps to build climate and disaster resilience into core investments across sectors, and offers unique financial solutions to better manage the contingent fiscal risks from disasters. Between 2006 and 2012, in East Asia and the Pacific, the World Bank financed disaster response projects totaling US$842 million and preventive projects amounting to US$2.1 million (GFDRR 2012). In addition, since 2007, the Global Facility for Disaster Reduction and Recovery (GFDRR) and other donors have been funding some 35 projects on DRM, amounting to more than US$32.7 million in the region (GFDRR 2012).[3]

World Bank DRM activities are part of a comprehensive framework. This framework focuses on five core areas of risk management: risk identification, risk reduction, emergency preparedness, financial protection, and sustainable recovery and reconstruction. Systematically addressing each core area, this report takes stock of the current situation of countries in East Asia and the Pacific, identifies the key challenges, and outlines priorities for policy makers to reduce risks and build resilience in the short, medium, and long terms.

Notes

1. ADB (2012) based on data from UN (2011).
2. Disasters Portfolio Database (data as of June 30, 2012). The database includes all projects with any activity related to disasters, although it excludes many activities that have a non–disaster-related purpose but that may also help to reduce the impact of disaster. Between 2006 and 2012, the World Bank committed an estimated US$11.7 billion to projects or project components related to DRM. In this time period, the Bank financed 113 disaster operations (US$7.9 billion) with ex ante activities (anticipating foreseeable disasters in the future) and 68 disaster operations (US$3.8 billion) with ex post activities (directly triggered by a disaster).
3. Internal projects database "RBMS" as of August 28, 2012. The figure includes multi-donor funds, donor-earmarked funds, and the Japan Social Development Trust Fund, but excludes the Japan Policy and Human Resources Development (PHRD) Fund and the Australian Agency for International Development (AusAID) East Asia Infrastructure Growth Fund (EAIIG) -funded projects.

References

ADB (Asian Development Bank). 2012. "Green Urbanization in Asia Key." Special chapter in *Indicators for Asia and the Pacific 2012*. Manila: Asian Development Bank.

GFDRR (Global Facility for Disaster Reduction and Recovery). 2012. Internal Disasters Portfolio and Projects Database. Washington, DC. https://www.gfdrr.org/gfdrr/node/44.

UN (United Nations). 2011. *World Population Prospects: The 2010 Revision*. Population Division of the Department of Economic and Social Affairs of the United Nations Secretariat, United Nations, New York. http://esa.un.org/wpp/index.htm.

Executive Summary

Where Are We Now?

Experiencing both recurrent small-scale events as well as devastating large-scale catastrophes, no other region in the world is affected by disasters as is East Asia and the Pacific. In the last decade, Ho Chi Minh City, Jakarta, Manila, and many other cities have been repeatedly hit by floods. In the last five years, Asia has experienced a large share of wide-scale natural catastrophes, including earth-quakes in the Tohoku region in 2011, Padang in 2009, and Wenchuan in 2008; typhoons in 2009 in the Lao People's Democratic Republic, the Philippines, and Vietnam; a cyclone in Myanmar in 2008; and large-scale floods in 2011 in Cambodia, Thailand, and the Philippines (figure ES.1). The year 2011 was the costliest year on record for natural disasters with cascading effects (Japan) and transboundary consequences (Thailand), adding up to US$380 billion in economic losses, almost doubling the 2005 record of US$262 billion.[1] In the first nine months in 2011, East Asia sustained about 80 percent of all disaster losses worldwide.

Growth of assets and population in harm's way is the single largest driver of disaster risk. Asia's urbanization is unique in terms of growth of population,

---

Box ES.1 Key Terms

Disaster prevention: Expresses the concept and intention to completely avoid potential adverse impacts through action taken in advance.

Disaster risk management: Processes for designing, implementing, and evaluating strategies, policies, and measures to improve the understanding of disaster risk, foster disaster risk reduction and transfer, and promote continuous improvement in disaster preparedness, response, and recovery practices, with the explicit purpose of increasing human security, well-being, quality of life, and sustainable development.

Disaster risk reduction: Denotes both a policy goal or objective, and the strategic and instrumental measures employed for anticipating future disaster risk; reducing existing exposure, hazard, or vulnerability; and improving resilience.

Figure ES.1 East Asia and the Pacific Disasters in Economic Losses in 2011

Great Tohoku earthquake March 2011	Thailand flooding September 2011	Christchurch earthquake February 2011	Australia flooding January 2011	Philippines flooding January 2011
The combined earthquake and tsunami in Japan left *15,853 dead, 6,013 injured,* and *3,286 missing*	Flooding in September and October caused *680 deaths* and affected more than *13 million* people	The Christchurch earthquake left more than *172 people dead*	Flooding along Australia's east coast affected more than *3 million* people and resulted in *22 deaths*	Flooding left *68 people dead, half a million* homeless, and *one million* affected
Damage US$210 billion	Damage US$46.5 billion	Damage US$12 billion	Damage US$9.8 billion	Damage US$4.5 billion

Sources: Adapted from UN OCHA 2011 with data from CRED-EMDAT; for Japan, Ministry of Finance 2012.

Figure ES.2 Asia's Unique Urbanization in Terms of Growth of Population, Cities, and Densities

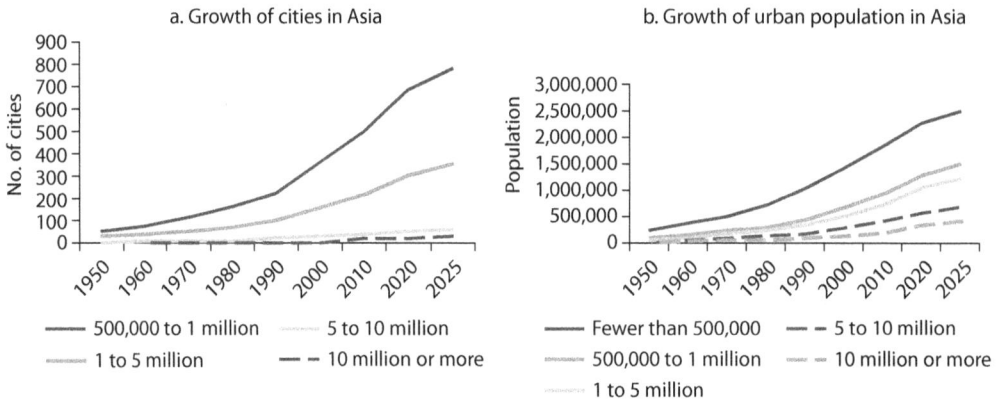

Source: Growth of cities by region. Data from UN 2011b.

cities, and densities (see figure ES.2 and appendix A for global comparison). From 1980 to 2010, Asia added more than one billion people to its cities—more than all other regions combined—and another billion dwellers will live in cities by 2040 (ADB 2012, iv). Urban Asia has high population densities and most of the world's megacities—by 2025, the number of megacities in Asia is expected to increase to 21 out of a global total of 37 (ADB 2012, iv). Growth of assets and megacities means that multi-billion-dollar disasters are becoming more widespread in the region. According to the Intergovernmental Panel on Climate Change Special Report on Extreme Events and the latest scientific evidence, "long-term trends in normalized losses have not been attributed to natural or anthropogenic climate change."[2] Unplanned or poorly planned rapid urbanization creates highly vulnerable communities, particularly through informal settlements and inadequate land management (IPCC 2012, 7).

Within East Asia and the Pacific, there is uneven capacity and readiness to invest in disaster risk management (DRM) (figure ES.3). The region includes developed countries with sophisticated institutions and instruments to effectively manage the risks of disasters (Australia, Japan, the Republic of Korea, Singapore, New Zealand), countries that have made considerable steps in mainstreaming DRM into development (China, Indonesia, the Philippines, Vietnam), and countries that face severe capacity and institutional constraints (Cambodia, Lao PDR, Mongolia, Republic of the Union of Myanmar, Pacific island countries). Small Pacific islands, Papua New Guinea, and Timor-Leste face serious challenges in their ability to recover from disasters. Significant capacity and funding gaps also exist between the central and local levels of government, as well as between rural and urban areas.

Low-income countries are unprepared and underfunded in their task to manage risk and lead recovery, and this is particularly true for local-level institutions. For example, globally, less than 20 percent of low- and lower-middle-income countries invest in land-use planning, less than 30 percent of low-income countries invest in landslide mitigation measures, and less than 50 percent of low-income countries invest in drainage infrastructure to mitigate flooding (figure ES.4) (UN 2011a, 87). Many countries in East Asia and the Pacific have made advances in DRM at the national level in formulating legislation and strategies,

Figure ES.3 Risk Governance Capacity and World Bank Country Classification by Income

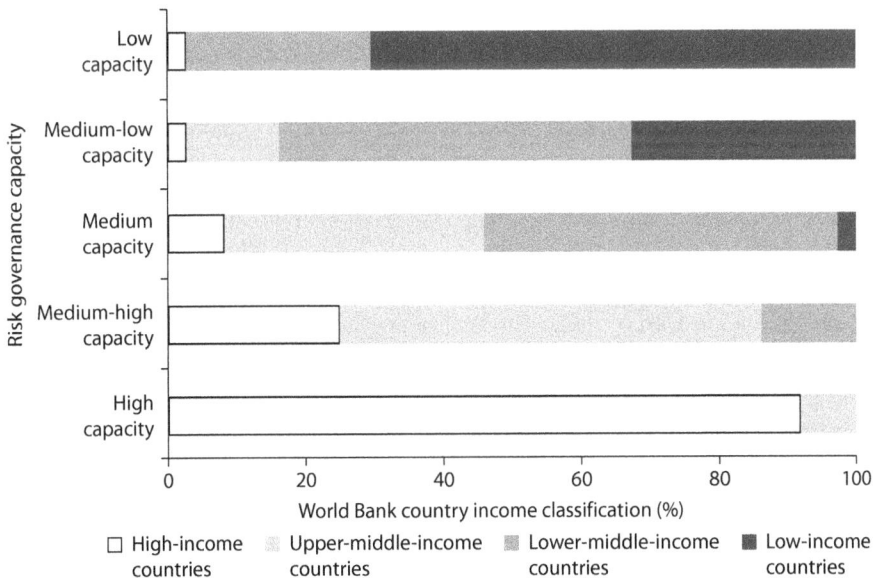

Source: UN 2011a, 7.
Note: World Bank income classification of countries: Low-income countries: Cambodia, Korea Dem. People's Rep., Republic of the Union of Myanmar; lower-middle-income countries: Fiji, Indonesia, Kiribati, Lao PDR, Marshall Islands, Micronesia, Fed. Sts., Mongolia, Papua New Guinea, Philippines, Samoa, Solomon Islands, Timor-Leste, Tonga, Vanuatu, Vietnam; upper-middle-income countries: American Samoa, China, Malaysia, Palau, Thailand, Tuvalu.

Figure ES.4 Underinvestment of Low- and Low-to-Middle-Income Countries in Risk Mitigation

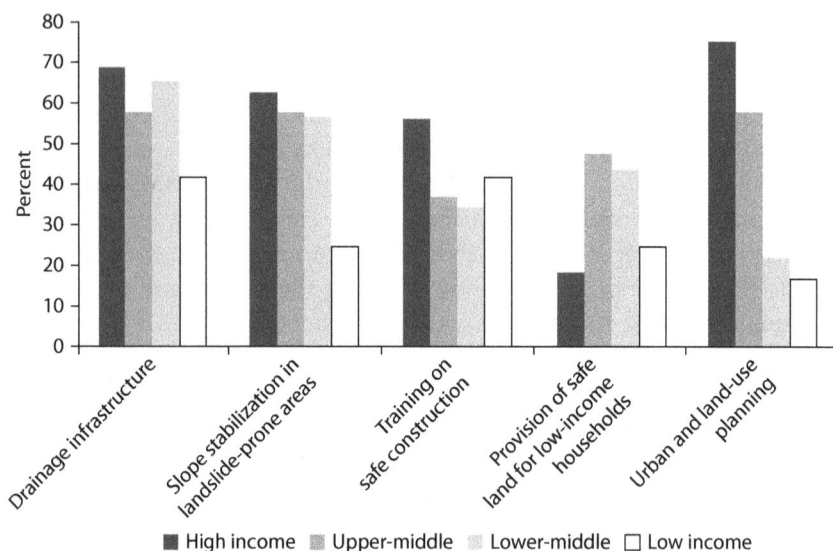

Sources: UN 2011a, 87; World Bank 2012b, 148.
Note: Percentage of countries reporting on DRM investments aimed at reducing urban risks.

but implementation remains a challenge. Local-level institutions lack the appropriate budgets, human resources, and technical capacity in their role as first responders and leaders in mainstreaming resilience into local-level investments.

Where Do We Want to Be?

On May 2, 2008, Cyclone Nargis, a Category 4 storm, made landfall in the Republic of the Union of Myanmar, killing an estimated 140,000 people.[3] On November 15, 2007, Cyclone Sidr, a Category 5 storm, made landfall in Bangladesh, a country at roughly the same level of per capita income as the Republic of the Union of Myanmar. The death count was 3,000.[4] How can we reduce human and physical costs of disasters?

Be prepared for the unexpected: Invest in disaster management and resilience. Every natural hazard does not necessarily need to turn into a disaster if societies at risk invest in disaster prevention, mitigation, and preparedness. Governments in the region and globally are under increasing pressure to protect their citizens and assets from harm caused by hazards. When people and assets are exposed to impacts of natural hazards, disaster risks should be considered in the design and implementation of development investments and services. Well-managed risks would save lives and protect livelihoods, resulting in lower capital losses from disasters (Hallegatte 2011). Even small investments in prevention represent savings in terms of avoided losses and reconstruction costs. Unfortunately, despite the benefits of prevention, the portion of disaster budgets spent on relief and repair still by far outweighs the fraction spent on disaster prevention.[5]

Make DRM part of core poverty alleviation and sustainable development efforts. Poverty and vulnerability to disasters are closely linked. The poor are disproportionately affected by disasters. Living in hazardous areas, the poor and vulnerable are more exposed to the impacts of disasters. Limited or no access to basic services, such as clean water, sanitation, and health services, exacerbates poor people's vulnerability to the impacts of disasters (World Bank 2000, 2012a). Additionally, they face constraints in their ability to respond and quickly recover their livelihoods because of lacking assets or resources, access to finance, and appropriate disaster response mechanisms. For example, a spatial assessment can show the exposure of low-income populations to multiple hazards (figure ES.5). Public investments can have a positive effect on countries' efforts to reduce poverty and promote sustainable economic growth, but where relevant, they need to be mindful of the impacts of disasters.

Prioritize actions to manage the risks from disasters based on informed decisions about the level of risk. A systematic assessment of the levels of risk can guide decisions whether to reduce, transfer, or manage disaster risks. Risk-informed urban planning and development provides a unique opportunity to

Figure ES.5 Patterns in Jakarta between Slum and Flood-Prone Areas

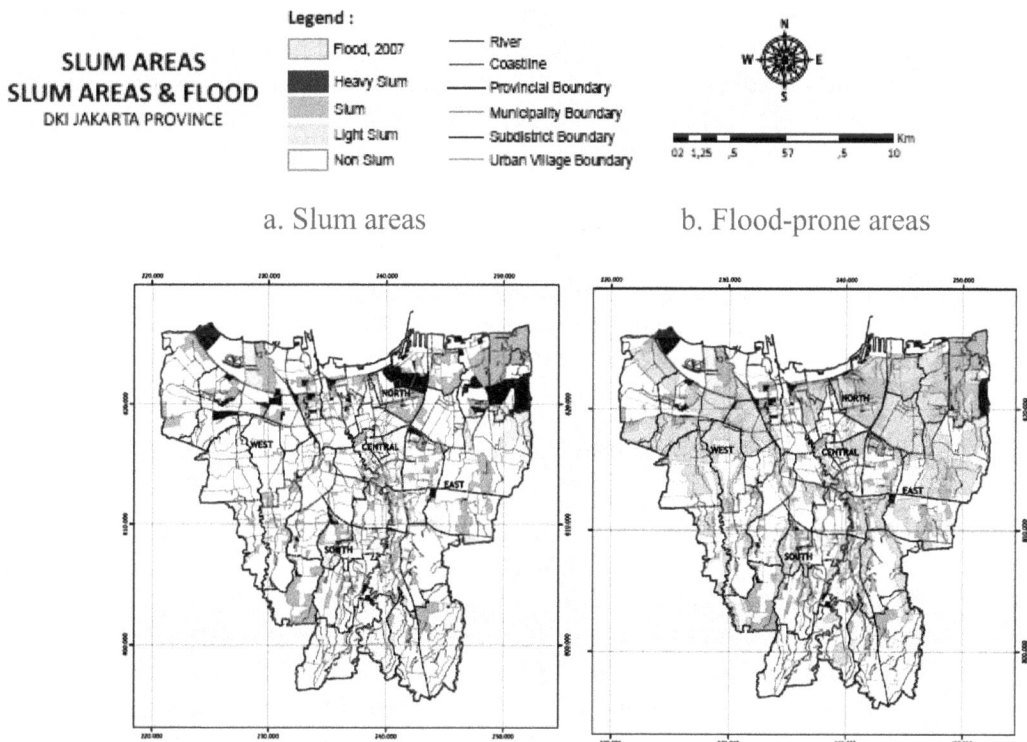

Source: World Bank 2011.

have a lasting positive impact on many lives. By addressing risk factors early on, developing participatory approaches, and combining them with regular infrastructure development, the negative growing risks linked with rapid urbanization can be minimized. For example, investing in "trunk infrastructure" to guide future city growth away from hazardous areas is an effective way for policy makers to reduce future disaster risk. Avoiding building schools or hospitals in flood-prone areas will reduce future unnecessary exposure to risks. Investing in reliable risk information and communicating such information in an effective manner to multiple stakeholders at the national, regional, city, and community levels helps to assess and communicate the socioeconomic and fiscal impacts of disasters and formulate effective DRM strategies. Sharing risk information between agencies across different levels and sectors and the public and the private sectors enables stakeholders to make better-informed decisions and strengthens the collective resilience of a community.

Manage residual risks and uncertainties by investing in the right balance of structural and nonstructural measures. In the context of complex networks and systems, a range of measures, including retrofitting critical infrastructure at risk, enforcing stringent building standards, strong institutional coordination and emergency procedures, and appropriate disaster risk-financing and insurance instruments, can help societies to live with disaster hazards without catastrophic consequences. Improved weather forecasting and early warning systems have led to striking results in reducing mortality risk. Although investments in both "gray" and "green" infrastructure are a crucial component of DRM, building the resilience of communities and households to deal with the shocks caused by disaster is equally important and particularly critical for poor and marginalized households. For example, community-based mechanisms can help to rapidly identify households most in need, groups recently pushed into poverty, and people missed by formal targeting systems, which often rely on pre-disaster data. To effectively manage the deep uncertainties related to disaster and climate change, planners are advised to focus on low-regret solutions, which are beneficial under a range of scenarios and fall under robust decision making. Good climate change risk management starts with a comprehensive and holistic approach to DRM.

What Needs to Be Done?

Managing risk and building resilience as part of an incremental and iterative process, depending on the country context, specific needs, and capacities, are essential. An assessment of the levels of risk, a cost-benefit analysis of available interventions, and an inventory of existing capacity and financial resources can guide decision makers in the prioritization of the recommendations presented in this report. Below is a list of policy actions that have proven to be effective in dealing with the existing and future risks linked to natural hazards and climate change (figure ES.6). Individual chapters of the report address the core elements of a comprehensive DRM strategy.

Figure ES.6 Making Informed Decisions to Manage Risks and Build Resilience

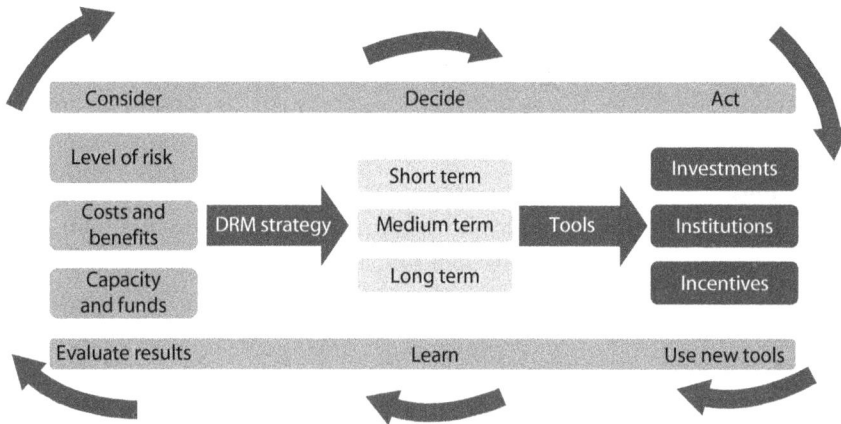

Source: Authors.

Short-Term: Early Warning Systems, Emergency Preparedness, and Risk Information

Investments

- Invest in hazard forecasting and hydrometeorological early warning systems, which can have a high positive cost-benefit ratio with immediate and significant payoffs.
- Invest in risk information, risk assessment, and catastrophic modeling systems.

Institutions

- Strengthen DRM legislation and arrangements and promote institutional coordination and capacity.
- Strengthen emergency preparedness and recovery planning, for example, through financial mechanisms to ensure rapid disbursement of funds in the aftermath of a disaster.

Incentives

- Promote disaster risk reduction in community-based development programs. Work together with communities and stakeholders to ensure that investments fit their needs.
- Promote sharing of risk information within relevant government agencies reaching different levels and sectors and the public and private sectors.

Medium- to Long-Term: The Right Balance between Structural and Nonstructural Measures

Investments

- Invest in structural and nonstructural measures based on the risk levels and costs versus benefits of the available options.

- Invest including "gray" concrete investments and cost-effective "green" infrastructure, such as establishing mangroves and wetland buffers and coastal restoration.
- Invest in expanding early warning systems based on real-time data and forecasting.

Institutions
- Develop a comprehensive disaster risk-financing strategy with ex ante and ex post instruments at the national and subnational levels.
- Develop (or build on existing) social protection systems and community-based interventions that can be rapidly scaled up in the event of a disaster and scaled down when no longer needed.

Incentives
- Promote sharing risk information between agencies across different levels and sectors and the public and the private sectors.
- Promote business emergency continuity planning by encouraging stakeholder consultation, implementation of standards, trainings, and simulation exercises.

Long-Term: Resilient Urban Development and Planning and Resilient Communities
Investments
- Invest in minimizing the consequences of urbanization through systematic use of risk assessments, risk-aware urban planning and development, and robust decisions that can take into account disaster and climate risks and uncertainties.
- Invest in working together with communities to build local-level resilience.

Institutions
- Enforce risk-based land-use planning and building codes.
- Enforce open access to risk information and tools.

Incentives
- Promote regional cooperation on DRM, especially in areas of risk information, weather and hazard forecasting, early warning systems, emergency preparedness, and risk financing and insurance.
- Promote private catastrophe risk insurance markets through public-private partnerships and the development of an enabling regulatory and risk market infrastructure.

How Can the World Bank Help?

Together with its partners and donors, the World Bank supports countries around the world in mainstreaming a comprehensive and holistic approach to

DRM into development. With its overarching mission to fight poverty, the World Bank's DRM efforts focus on building resilient communities. In East Asia and the Pacific, the World Bank supports a range of low- to upper-high-income countries in developing effective ex ante and ex post risk management measures. Paying close attention to countries' individual needs, the World Bank provides analytical and advisory services, helps to build climate and disaster resilience into core investments across sectors, and offers unique financial solutions to better manage the contingent fiscal risks from disasters.

World Bank engagement is based on a comprehensive risk management framework focusing on five core areas: risk identification, risk reduction, emergency preparedness, financial resilience, and sustainable recovery and reconstruction, as illustrated in figure ES.7 (see also Appendix H). This framework supports the implementation of the Hyogo Framework on Actions, the international agreement of disaster risk reduction, and provides an effective way for countries to coordinate, harmonize, and leverage government- and donor-led activities. The World Bank Global Facility for Disaster Reduction and Recovery (GFDRR) helps to leverage partnerships on the ground and transfer global knowledge of DRM.

1. **Risk identification:** Technical assistance with creation, retention, and sharing of vital risk information, through nation- and region-wide risk assessments, analytical tools, and innovative ways of sharing vital information with key stakeholders and populations at risk.
2. **Risk reduction:** Technical assistance and investment lending for structural measures, provision and improvements of basic services, retrofitting of critical infrastructure and nonstructural measures, risk-based land-use planning and

Figure ES.7 World Bank's DRM Framework and Examples of Engagements in East Asia and the Pacific

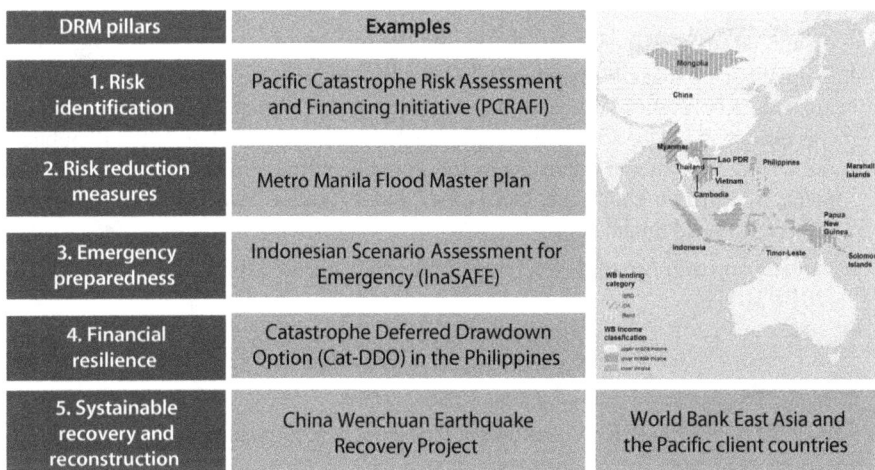

DRM pillars	Examples	
1. Risk identification	Pacific Catastrophe Risk Assessment and Financing Initiative (PCRAFI)	
2. Risk reduction measures	Metro Manila Flood Master Plan	
3. Emergency preparedness	Indonesian Scenario Assessment for Emergency (InaSAFE)	
4. Financial resilience	Catastrophe Deferred Drawdown Option (Cat-DDO) in the Philippines	
5. Systainable recovery and reconstruction	China Wenchuan Earthquake Recovery Project	World Bank East Asia and the Pacific client countries

Source: Authors.

urban development planning, provision of social funds and safety nets, and strengthening risk awareness and preparedness through community-driven and community-based programs.

3. **Emergency preparedness:** Technical assistance and investment lending for early warning and monitoring systems, emergency response planning, and risk communication.

4. **Financial resilience:** Technical assistance and ex ante and ex post funding mechanisms and services as part of a comprehensive disaster risk financing and insurance strategy.

5. **Sustainable recovery and reconstruction:** Technical assistance, ex ante and ex post funding mechanisms for a quick recovery, and improved institutional planning inclusive of the needs of the most vulnerable populations.

Effective DRM requires a strong partnership between multiple stakeholders. With the right investments, institutions, and incentives, the growth of cities and economies can be channeled as a tremendous positive force for development. Natural hazards are inevitable, but through a better understanding and communication of risk, more balanced risk reduction measures, and more effective risk transfer and management actions, the impacts of natural hazards can be reduced significantly, saving lives, preventing losses, and safeguarding development.

Notes

1. Figures from Munich Re NatCatService. Note that estimates differ: EM-DAT estimates damage losses at US$366.1 billion for natural disasters that occurred in 2011. The Swiss Re estimate is US$370 billion.

2. Normalized for asset and population growth, there is no climate change signal in damages from extreme weather events for the foreseeable future. See Barthel and Neumayer (2011), "At a global scale no significant trend is discernible. Similarly, we do not find a significant trend if we constrain our analysis to non-geophysical disasters in developed countries," and Pielke et al. (2008).

3. Includes missing population; ASEAN (2009).

4. Estimates vary, suggesting up to 10,000 deaths. Although this is still high, early warning and early action to evacuate people and shelters reduced the death toll. Damage was estimated at US$450 million in Bangladesh and US$10 billion in the Republic of the Union of Myanmar according to the Republic of the Union of Myanmar government. Read more about the Bangladesh 2007 Cyclone Recovery and Restoration Project at World Bank (2008).

5. For data on Latin America, see, for example, de la Fuente (2009) cited in World Bank (2010, 107).

References

ASEAN. 2009. *Fourth ASEAN State of the Environment Report.* Jakarta.

Barthel, F., and E. Neumayer. 2011. "Normalizing Economic Loss from Natural Disasters: A Global Analysis." *Global Environmental Change* 21 (1): 13–24.

Hallegatte, S. 2011. "How Economic Growth and Rational Decisions Can Make Disaster Losses Grow Faster than Wealth." Policy Research Working Paper 5617, World Bank, Washington, DC.

IPCC. 2012. *Managing the Risks of Extreme Events and Disasters to Advance Climate Change Adaptation.* Special Report of Working Groups I and II of the Intergovernmental Panel on Climate Change. Edited by C. B. Field, V. Barros, T. F. Stocker, D. Qin, D. J. Dokken, K. L. Ebi, M. D. Mastrandrea, K. J. Mach, G.-K. Plattner, S. K. Allen, M. Tignor, and P. M. Midgley. Cambridge, UK: Cambridge University Press.

Japan, Ministry of Finance. 2012. Great East Japan Earthquake website. Accessed October 2012. http://www.mofa.go.jp/j_info/visit/incidents/index2.html.

Pielke, R. A., Jr., J. Gratz, C. W. Landsea, D. Collins, M. A. Saunders, and R. Musulin. 2008. "Normalized Hurricane Damage in the United States: 1900–2005." *Natural Hazards Review* 9 (1): 29–42. http://sciencepolicy.colorado.edu/admin/publication_files/resource-2476-2008.02.pdf.

UN (United Nations) OCHA. 2011. *Humanitarian Funding Update First Quarter 2011.* Online brief. Accessed September 2012. http://ochanet.unocha.org/p/Documents/Asia%20Pacific%20Funding%20Update,%20First%20Quarter%202011.pdf.

UN ISDR (UN International Strategy for Disaster Reduction). 2009. *Terminology on Disaster Risk Reduction.* UNISDR: Geneva, Switzerland. http://www.unisdr.org/files/7817_UNISDRTerminologyEnglish.pdf.

World Bank. 2000. *Managing Disaster Risk in Emerging Economies.* Edited by A. Kreimer and M. Arnold. Washington, DC: World Bank.

———. 2008. "World Bank Provides Support to Cyclone Affected Areas in Bangladesh." Press Release, November 6. http://go.worldbank.org/53AW0ZTGE0.

———. 2010. *Natural Hazards, Unnatural Disaster: The Economics of Effective Prevention.* Washington, DC: World Bank.

———. 2011. "Jakarta: Urban Challenges in a Changing Climate." Working Paper 65018, World Bank Indonesia, Jakarta. http://documents.worldbank.org/curated/en/2011/01/15310553/jakarta-urban-challenges-changing-climate.

———. 2012a. *Climate Change, Disaster Risk, and the Urban Poor: Cities Building Resilience for a Changing World.* Edited by J. Baker. Washington, DC: World Bank.

———. 2012b. *Inclusive Green Growth: The Pathway to Sustainable Development.* Washington, DC.

CHAPTER 1

Managing Risks in East Asia and the Pacific: An Agenda for Action

Key Messages for Policy Makers

- Rapid urbanization, the growth of urban populations and assets in combination with poorly or unplanned development, is the main driver of the cost of disasters in the region.
- The costs of inaction can be very high, in terms of both lives lost as well as economic damages.
- A comprehensive disaster risk management (DRM) framework offers practical opportunities for targeted policy action and investments, stretching across sectors and jurisdictions and reaching all the way to the communities and the most vulnerable populations.
- Managing disaster risk and building resilience is an incremental process. An assessment of the levels of risk, a cost-benefit analysis of available interventions, and an inventory of existing capacity and financial resources can guide the prioritization of actions.
- To deal with the risks and deep uncertainties linked to natural hazards and climate change, planners are advised to focus on robust, low-regret solutions that can bring benefits under many different scenarios.

Where Are We Now?

East Asia and the Pacific is frequently affected by small and recurrent disasters as well as rare high-impact events (figure 1.1). Home to 59 percent of the world's population and covering half the earth's surface area, East Asia and the Pacific has experienced more than 70 percent of the world's disasters and 82 percent of disaster fatalities. In the past year (2011–12), the east Japan earthquake and tsunami, large-scale floods in Thailand, and a tropical storm in

This chapter was written by Abhas Jha and Zuzana Stanton-Geddes with input from Liana Zanarisoa Razafindrazay.

Figure 1.1 Impact of Natural Disasters in East Asia and the Pacific in the Last 30 Years

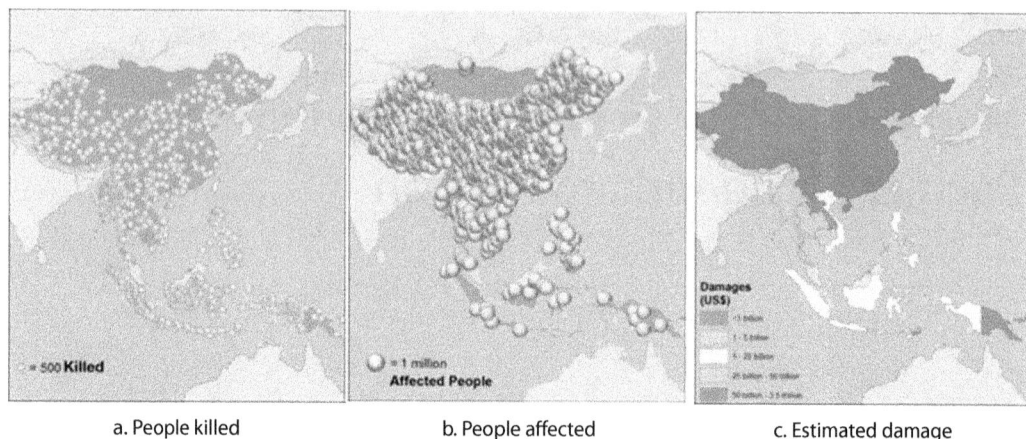

a. People killed   b. People affected   c. Estimated damage

Country	People killed	People affected	Estimated damage (US$, thousands)
Cambodia	1,426	18,319,666	1,057,110
China	154,602	2,974,972,174	356,292,317
Fiji	221	1,152,658	529,733
Indonesia	188,610	21,686,240	14,356,527
Kiribati	0	84,085	0
Lao PDR	207	5,465,868	429,779
Marshall Islands	0	6,818	0
Mongolia	235	3,259,092	152,364
Rep. of the Union of Myanmar	139,385	3,949,129	4,707,943
Papua New Guinea	3,008	1,346,645	178,253
Philippines	34,383	129,556,382	8,194,581
Samoa	174	290,585	676,600
Solomon Islands	181	298,682	20,000
Thailand	12,781	80,795,502	46,671,747
Timor-Leste	5	13,571	0
Vanuatu	212	283,529	205,000
Vietnam	15,689	74,944,401	8,629,252
East Asia and the Pacific total	551,119	3,316,425,027	442,101,206

Source: EM-DAT 2012 data for years 1980–2011.
Note: These figures depict information only from the listed countries. Total damage loss is estimated at US$453 billion for the years 1980–2011, based on data from EM-DAT.

the Philippines are tragic reminders of the devastation, economic damage, and loss of human life caused by disasters, conveying important lessons to urban DRM practitioners. In relative terms, the Pacific small island developing states are among those most affected globally, with average annualized losses estimated for Vanuatu and Tonga at 6.6 and 4.4 percent of gross domestic product (GDP), respectively. Fiji's main entry port was twice up to 4 meters under water during record floods in 2009 and 2012.

The rising cost of disasters is eroding the hard-won gains of economic development. Statistics show that, even when adjusted for inflation, the losses caused by natural catastrophes have been increasing dramatically and at an ever-quickening pace since 1950. In the period between 1990 and 1999, the costs of disasters in constant dollars were more than 15 times higher than in the 1950–59 period. Figure 1.2 shows the average damages in billions of U.S. dollars in the years 2000–08. The year 2011 was the costliest on record for disasters with economic losses exceeding the previous record of US$262 billion in 2005 by nearly 50 percent (Guha-Sapir et al. 2012). Asia was the continent most hit by disasters in 2011: 44 percent of disasters, accounting for 83.6 percent of global disaster victims, and bearing 75.4 percent of total damages (Guha-Sapir et al. 2012).

East Asia and the Pacific faces disconcerting emerging trends, namely, urbanization of disasters, increasing frequency of urban flooding, rising complexity of disasters, as well as the cross-regional impacts of disasters.

The risks in East Asia and the Pacific will continue to rise as the population and wealth in cities at all scales increases. The phenomenal urbanization in Asia is largely driven by rapid economic growth. Developing countries will absorb most of future urban growth and development, especially small and medium cities (figure 1.3). From 1980 to 2010, Asia added more than one billion people to its cities—more than all other regions combined—and another billion inhabitants are expected to live in urban areas by 2040 (ADB 2012, iv). Concentrating businesses, knowledge and technology, and an educated labor force, cities are traditional drivers of development. However, with a higher concentration of people and assets, urban areas are also particularly vulnerable. Historically many urban centers have been located in hazardous zones, for example, at sites of agricultural surplus such as fertile volcanic soils or along major trade and transportation

Figure 1.2 Weather and Climate-Related Disasters and Regional Average Impacts, 2000–08
US$, billions

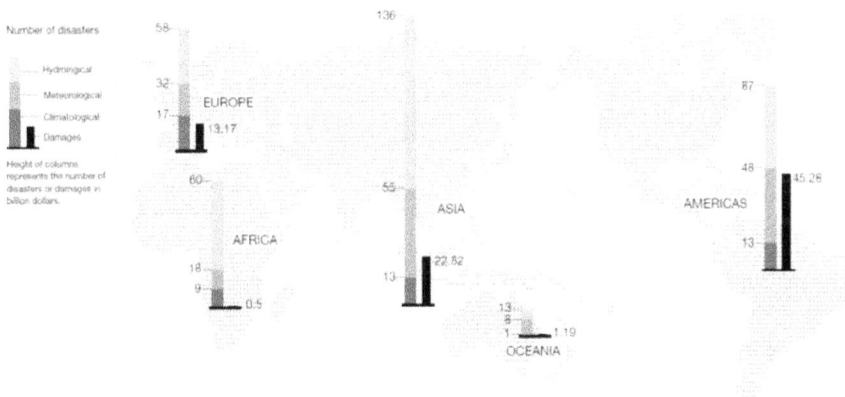

Source: Climate and Development Knowledge Network 2012.

Strong, Safe, and Resilient • http://dx.doi.org/10.1596/978-0-8213-9805-0

Figure 1.3 Growing Assets in Asia

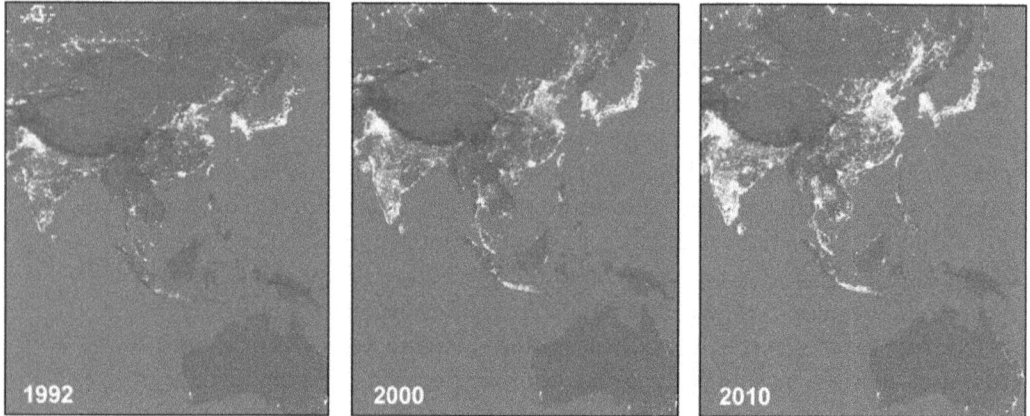

routes such as coasts and river systems that are prone to flooding and coastal erosion (Dilley et al. 2005; see also Hallegatte 2011) or are located on seismic faults.

The increase in concentrations of people and growth of assets in hazardous areas is the single largest driver of disaster risk and greatest challenge for managing disaster risks. According to the Intergovernmental Panel on Climate Change Special Report on Extreme Events and the latest scientific evidence, "long-term trends in normalized losses have not been attributed to natural or anthropogenic climate change." Although climate change will have an increasing impact on growth in the future, the main reason for the rising cost of disasters has been the concentration of the world's population and economic activity in vulnerable locations near earthquake faults, on subsiding river deltas, and along tropical coastal zones (see also appendix C for details). Normalizing losses for growth of assets over time, there are "no significant upward trends in normalized disaster damage over the period 1980–2009 globally, regionally, for specific disasters or for specific disasters in specific regions" (figure 1.4) (Barthel and Neumayer 2011a, 225).[1]

The vulnerability to flooding will increase in Asia, with a projected 410 million urban Asians at risk of coastal flooding by 2025 (ADB 2012; World Bank 2012a). East Asia and the Pacific accounts for about 40 percent of the total number of floods worldwide over the past 30 years. Urban flooding is becoming increasingly costly as low- and middle-income countries transition to largely urban societies. Flash floods and seasonal river flooding occurs throughout the region with a risk of storm surges along many coastlines. Tropical cyclones are the most costly meteorological disasters affecting East Asia and the Pacific with, on average, 27 tropical cyclones affecting some part of the region each year (Chan 2008).[2] This is a major issue for national disaster management agencies, with the

Figure 1.4 Normalizing Losses from Nongeophysical Disasters in South and East Asia and Pacific Countries with Different Methodologies

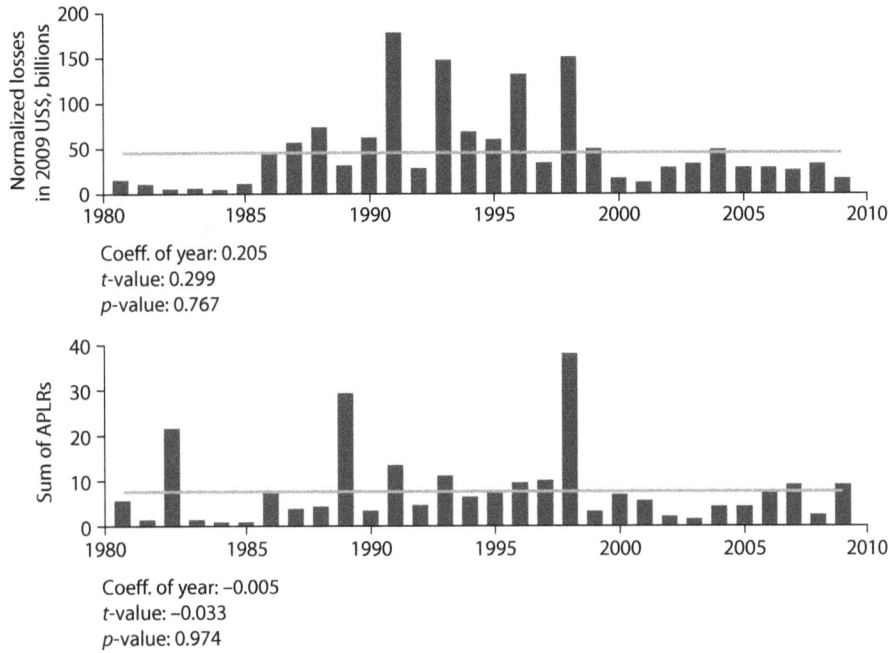

Coeff. of year: 0.205
t-value: 0.299
p-value: 0.767

Coeff. of year: −0.005
t-value: −0.033
p-value: 0.974

Source: Barthel and Neumayer 2011b.
Note: Based on 3,858 disasters. Losses normalized with conventional approach (top) and alternative approach (bottom).
APLR = actual-to-potential-loss ratio.

impact usually highest in the poorest neighborhoods, which are often the most vulnerable and least prepared (World Bank 2012b).

With the growing size of cities and complex product and supply networks, many countries face deep uncertainty over cascading disasters with cross-boundary and cross-regional impacts. In Thailand, damage and disruption reached well beyond the scope of floodwaters. Total economic damages were estimated at US$46.5 billion, with more than 90 percent borne by the private sector (World Bank 2012d). As the impact of the disaster spread through industrial supply chains, losses were felt across production networks in Asia and globally. Recent large-scale disasters, including the Thai floods and the east Japan earthquake and tsunami, are a reminder that, despite best efforts, accidents are inevitable (Perrow 2011). In the complex, tightly networked systems of modern society, accidents are "normal" as a single-event failure or a number of disconnected failures with devastating consequences (Perrow 2011).

Where Do We Want to Be?

The ultimate objective of DRM efforts is to reduce risk, manage residual risk and uncertainties, and build resilient communities. There are encouraging developments leading up to this goal, but more needs to be done in the region, especially in cities, such as Ho Chi Minh City, Jakarta, Manila, Seoul, and other fast-growing dense cities in low-lying areas. In the last few decades, risk reduction efforts have succeeded in reducing the global death toll of natural disasters, despite the world's growing population, through improved early warning, more stringent building codes, and better contingency planning. Building upon these efforts, more can be done to reduce the costs of disasters. Experience from the recent disaster in the Tohoku region in Japan demonstrated that investments in prevention represent a savings in terms of avoided losses and reconstruction costs, and even small investments can have large benefits (box 1.1, see also World Bank 2010a and 2010b).

Box 1.1 Lessons from the Tohoku Earthquake

The east Japan earthquake and tsunami of March 2011 was one of the most powerful earthquakes ever to have hit Japan, and one of the five most powerful earthquakes in the world since modern record-keeping began in 1900. The Tohoku disaster is a testament to Japan's careful investments over many years in seismic safety and tsunami early warning systems as well as a reminder that although even the best of preparations pay off, they cannot fully insulate people and assets from the impact of disasters.

The ongoing Japan–World Bank collaborative project "Learning from Mega Disasters: A Program of Knowledge Sharing and Knowledge Exchange" aims to capture the key lessons emerging from Japan's resilience and its response and recovery efforts. A series of *Knowledge Notes* and the *Sendai Report* take stock of current DRM knowledge and make recommendations for building resilience into development. Japan and the World Bank cohosted the Sendai Dialogue as part of the program for the 2012 IMF–World Bank Group Annual Meetings to highlight lessons learned from the Tohoku disaster and adopt them as guidance for comprehensive DRM in at-risk countries around the world.

Select lessons for countries in East Asia and the Pacific include the following:

- Extreme disasters underscore the need for a **holistic approach to DRM** because single-sector planning cannot address the complexity of problems posed by natural hazards.
- **Preventive investments** pay off over the long term, but one must expect the unexpected because natural hazards can never be completely eliminated. This necessitates integrated disaster risk reduction that balances structural and nonstructural solutions.
- **Assessing risks and communicating** them clearly and widely among communities helps residents to make timely decisions to protect themselves.
- **Coordination mechanisms** must be developed and tested in normal times so that they are ready for use in an emergency.
- **Social safety nets** are needed in times of both emergencies and reconstruction to protect and engage with vulnerable groups.
- **Importance of "South-South" and peer-to-peer assistance:** Municipalities across Tohoku sent DRM and planning experts to the affected districts.

Sources: Based on World Bank 2012e, 2012f.

In the context of rapid urbanization and climate change uncertainties, robust decision-making processes, building redundancy, and community resilience are vital. Rather than trying to find the optimal protection solutions, we need to adopt a robust approach to uncertainty and unknown risks that incorporates a greater degree of flexibility into the mitigation design measures and that takes into account potential weak spots and failure. This is particularly true for complex systems and networks in which accidents and failure are to an extent inevitable (Perrow 2011). Box 1.2 gives insights in how to deal with uncertainty in increasingly more complex and intertwined societies.

When dealing with complex systems, preparing for failure under multiple scenarios and following low-regret solutions can help to reduce disaster impact. "Infrastructure robustness and redundancy are critical to maintaining the functions of the economic system after disasters, especially in urban environments, where the failure of one component (such as electricity, transport, water, or sanitation) can paralyze activities" (World Bank 2012c). This robust

---

Box 1.2 Approaches to Dealing with Complex Failures and Uncertainty

In the face of increasingly more complex infrastructure systems, what risk management strategies are necessary to prevent cascading failures?

Traditionally, complex infrastructure systems have been designed to resist the loads imparted. Yale sociologist Charles Perrow uses the term **vulnerability of complexity** to refer to failures repeatedly created at the intersections of our interdependent and highly sophisticated transportation, electric power, and telecommunications systems. Their interdependence make these core infrastructures vulnerable both to failures in each other and in the information systems and software that support their operations.

Richard Little argues for a more adaptive approach to infrastructure design. More adaptive approaches will not stop failures from occurring, but they do give us more options for reducing the consequences. At the same time, we need to move our thinking from an assumption that we can keep ourselves safe from all the extreme effects of nature and technology. We must do the best job possible, but a limit certainly exists to the ability of governments and institutions as to what they can do, and we should be more realistic about these limitations and communicate this reality to the public.

The 2009 financial crisis highlighted the interconnectivity and fragility of the complex banking system, when put under extreme pressure. Andrew Haldane, the Executive Director for Financial Stability and member of the Financial Policy Committee at the Bank of England, presents evidence from a range of real-world settings to demonstrate that decision making in a complex environment can benefit from the use of simple decision rules of thumb. He argues that complex rules often have punitively high costs of information collection and processing, rely on "overfitted" models that yield unreliable predictions, and can induce defensive behavior by causing people to manage to the rules. He says that "simplicity rather than complexity may be better capable of solving these robustness problems."

box continues next page

Box 1.2 Approaches to Dealing with Complex Failures and Uncertainty *(continued)*

In practice, cities can guide their actions by "safety margins strategies." Hallegatte notes that in Copenhagen, Denmark, to calibrate drainage infrastructure, "water managers use run-off figures that are 70 percent larger than their current level." The higher margin is set to buffer potential urban population growth as well as to cope with future climate change impacts, which could increase precipitation levels. The 70 percent increase is not the result of a precise calibration, given that this would be impossible to calculate because of uncertainty linked to climate change impacts. Based on information available, the margin "is thought to be large enough to cope with almost any possible climate change during this century." A lesson for other cities would be to be rather overly pessimistic in the design phase, because it is "inexpensive to implement a drainage system able to cope with increased precipitation" whereas "modifying the system after it has been built is difficult and expensive" (Hallegatte 2009).

Sources: Haldane and Madouros 2012; Hallegatte 2009; Little 2009; World Bank 2011.

---

approach can help authorities avoid being "locked in" to finance large-scale investments that might prove obsolete with change in future risks. This approach underlines the fact that good climate change risk management should start with a comprehensive and holistic approach to DRM and with measures that make sense under any future scenarios, such as reducing social vulnerability, investing in early warning systems, and promoting core urban planning. Figure 1.5 illustrates consideration of options based on their cost-benefit ratios in the context of deep uncertainty. In the Netherlands, to be able to make robust decisions, the decision makers use a decision tree that highlights points where decisions need to be taken, and where multiple solutions are possible (figure 1.6).

What Needs to Be Done?

Countries in East Asia and the Pacific need to continue to strengthen their efforts in reducing and managing disaster risks. An assessment of the levels of risk, a cost-benefit analysis of available interventions, and an inventory of existing capacity and financial resources can guide decision makers in the prioritization of actions. Understanding the level of risk, making conscious decisions whether to accept a portion or risk, reduce, transfer, or manage risks, and then choosing available instruments are part of a necessary informed decision-making process. DRM is an iterative process, in which lessons learned and new technologies can help us to better adapt to changed circumstances (figure 1.7).

The following chapters of this report focus on specific areas of this framework and provide concrete suggestions for action. Institutional and capacity building is a cross-cutting priority to support the policy and implementation framework for DRM, knowledge transfer toward and within the region, local-level capacity, community engagement, and fostering partnerships within the

Figure 1.5 Robustness to Climate Change Uncertainties

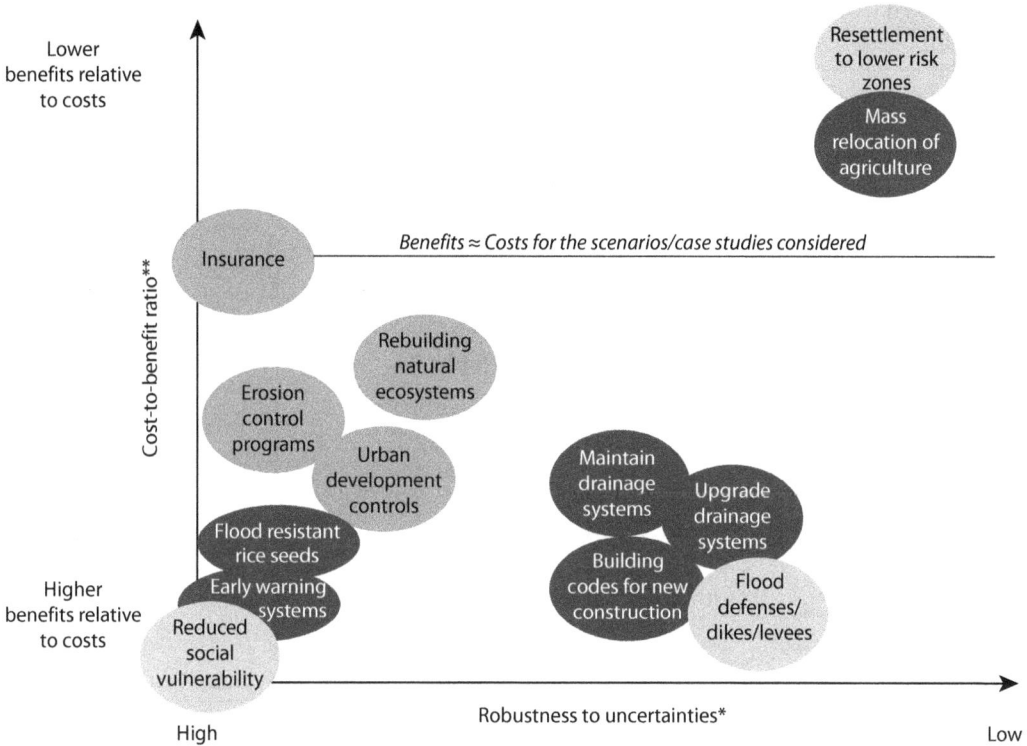

Source: Ranger and Garbett-Shiels 2011 cited in World Bank 2012a.
Note: Figure based on case studies of Guyana and Mozambique. * = The placement of measures on this axis is based on general flexibility/sunk-cost considerations but will vary depending on the individual case. ** = This will vary by case, so is only indicative. It is based on present-day climate and takes into account co-benefits.

countries, regionally and internationally. Equally, building social resilience through a deeper understanding of the social impacts of disasters can lead to more effective preventive actions, more responsive and cost-effective rehabilitation programs, and overall faster recovery and reconstruction. Regional cooperation on DRM can bring many benefits, especially in areas of risk information, weather and hazard forecasting, early warning systems, emergency preparedness, and risk financing and insurance.

Risk Identification
- **Develop risk information and modeling systems** to assess the economic and fiscal impact of disasters and include those risks in overall fiscal risk management.
- **Share vital risk information** within relevant government agencies reaching different levels and sectors, as well as communities and the private sector.
- **Open access to information and enhance the capacity of stakeholders** to use the available risk information and tools.

Figure 1.6 Formulating an Adaptive Strategy: Experience from the Netherlands

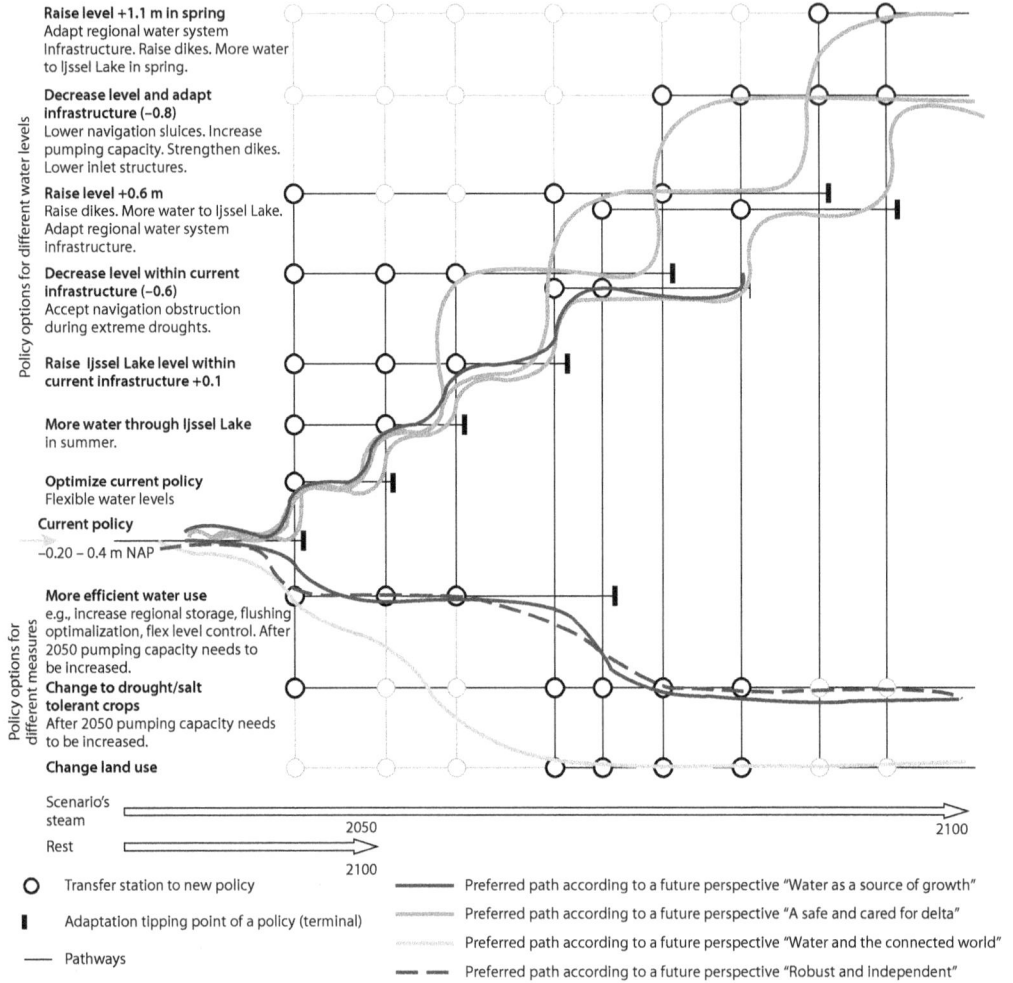

Source: Deltares 2012.

Risk Reduction

- **Provide strong central coordination and local capacity building:** Risk reduction is multidisciplinary and needs to be mainstreamed into all sectors at risk.
- **Integrate risk-based methods into cost-benefit approaches** to quantify the economic consequences of climate change and disaster impacts, and focus on low-regret solutions with high benefits under multiple scenarios.
- **Minimize the consequences of urbanization through the use of risk assessments** in decision making and risk-aware urban planning.
- **Promote balanced investments for structural and nonstructural measures** by promoting institutional arrangements, land-use regulations, natural buffers, and risk awareness along infrastructure investments.

Figure 1.7 Informed Decision-Making Process to Manage Risks and Build Resilience

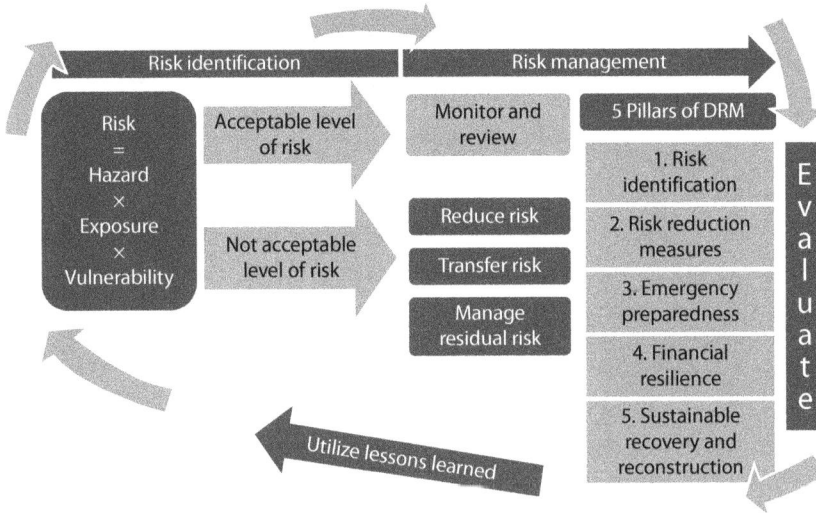

Source: Authors.

- **Promote disaster risk reduction in community-based development programs.** Invest and work together with communities and stakeholders to ensure that proposed medium- and long-term investments are responsive to their needs.
- **Build on existing (or develop) social protection systems and community-based interventions** that can be rapidly scaled up in the event of a disaster, and scaled down when no longer needed.

Emergency Preparedness
- **Invest in hazard forecasting and end-to-end early warning systems,** because these are no-regret investments that should be made as soon as possible.
- **Facilitate business emergency continuity planning** to address risks facing product and supply chains, by encouraging stakeholder consultation, implementation of standards, training, and simulation exercises.

Financial Protection
- **Develop disaster risk-financing and insurance strategies** at the national and subnational levels to manage potential budget volatility associated with natural hazards and provide insurance coverage against disasters for key public assets.
- **Establish national disaster funds** as a financial mechanism to ensure the rapid disbursement and execution of funds in the aftermath of a disaster.
- **Support the development of risk insurance infrastructure** to assist the development of a cost-effective, affordable, and sustainable insurance market and facilitate disaster risk pooling, creating a larger, more diversified

portfolio that should lead to lower reinsurance prices and reduced transaction costs.

- **Promote private catastrophe risk insurance markets** through public-private partnerships and the development of an enabling regulatory and risk market infrastructure that controls insurers' exposure accumulations to catastrophe risk using a risk-based capital approach.

Sustainable Recovery and Reconstruction

- **Strengthen emergency preparedness and recovery planning,** bearing in mind the needs of the most vulnerable populations.
- **Start recovery programs with an understanding of local realities** and channel appropriate resources to help support the priorities and needs of affected communities.
- **Encourage local capacity building and stakeholder participation** to achieve faster disaster recovery and more resilient reconstruction.

How Can the World Bank Help?

The World Bank supports countries around the world in mainstreaming a holistic approach to DRM into development. Historically the World Bank is one of the largest institutions financing recovery, reconstruction, and, increasingly, disaster risk prevention. With its overarching mission to fight poverty, the World Bank's DRM efforts focus on building resilient communities. The World Bank supports sustainable economic growth that is efficient, clean, and resilient—efficient in its use of natural resources, clean in that it minimizes pollution and environmental impacts, and resilient by accounting for natural hazards and the role of environmental management and natural capital in preventing physical disasters (World Bank 2012c). Between fiscal years 2006 and 2012, the International Bank for Reconstruction and Development (IBRD) and the International Development Association (IDA) committed an estimated US$11.7 billion to projects or project components related to DRM.[3] During the last seven years, the Bank has financed 113 disaster operations (US$7.9 billion) with ex ante activities (anticipating foreseeable disasters in the future) and 68 disaster operations (US$3.8 billion) with ex post activities (directly triggered by a disaster) (GFDRR 2012).[4]

In East Asia and the Pacific, the World Bank supports a range of low- to upper-high-income countries in developing effective ex ante and ex post risk management measures.[5] Paying close attention to individual country context, the World Bank provides advisory and analytical services, global expertise, capacity building and technical assistance, leverages partnerships, links communities of practice and networks, and provides targeted investment as well as programmatic lending. Between 2006 and 2012, in East Asia and the Pacific, the World Bank financed disaster response projects totaling US$842 million and preventive projects amounting to US$2.1 million (GFDRR 2012). In addition, since 2007,

Box 1.3 The Global Facility for Disaster Reduction and Recovery

The World Bank hosts the Global Facility for Disaster Reduction and Recovery (GFDRR), a growing partnership of 41 countries and eight international organizations, including the United Nations and the European Union. GFDRR was established in 2006 to assist countries to reduce disaster losses by 2015, in response to the Hyogo Framework for Action (HFA) agreed to at the World Conference on Disaster Reduction. GFDRR has worked with the World Bank to move from a reactive approach to disasters to become more strategic in supporting the reduction of disaster risk.

GFDRR has leveraged the World Bank's role, leadership, and performance on global knowledge creation, innovation, and partnerships in DRM. It has increased the World Bank's capacity and strategic planning to provide assistance to integrate DRM and climate adaptation in country development strategies, undertake timely post-disaster needs assessment (PDNA), and support country capacity building.

Source: World Bank 2012f.

GFDRR (see box 1.3) and other donors have been funding about 35 DRM projects, amounting to more than US$32.7 million in East Asia and the Pacific (GFDRR 2012).[6] In the region, the World Bank has ongoing partnerships with the Asian Development Bank (ADB), Association of Southeast Asian Nations (ASEAN), Asian Disaster Preparedness Center (ADPC), Asian Disaster Reduction Center (ADRC), Australian Agency for International Development (AusAID), Australia-Indonesia Facility for Disaster Reduction (AIFDR), Japan Aerospace Exploration Agency (JAXA), Japan International Cooperation Agency (JICA), the Republic of Korea National Emergency Management Agency (NEMA), Nanyang Technological University's Institute of Catastrophe Risk Management, Applied Geo-Science and Technology Division of the Secretariat of the Pacific Community (SOPAC), United Nations International Strategy for Disaster Reduction (UNISDR), World Meteorological Organization (WMO), and others.

World Bank DRM activities are part of a comprehensive framework focusing on several core areas: risk identification, risk reduction, emergency preparedness, financial protection, and sustainable recovery and reconstruction, as illustrated in figure ES.7 and figure 1.7. They follow the World Bank Group's strategic priorities by supporting resilient growth, improving livelihoods, good governance, regional partnerships, and disaster and climate change preparedness and mitigation, and implementing the international framework on disaster risk reduction, the Hyogo Framework for Action (HFA).[7] Box 1.4 provides an example of one country taking concrete steps in developing a comprehensive DRM framework.

Box 1.4 Strengthening the Philippines' Resilience to Disasters

Challenge: The Philippines is highly exposed to disasters, with an estimated 74 percent of the population vulnerable to natural hazards. Catastrophic disasters occurring once every 200 years could result in contingent liability in excess of 8 percent in the Philippines, totaling 18 percent or more of total public expenditure (World Bank 2012g). In October 2009, the Philippines was hit by the devastating Tropical Storm Ondoy (Ketsana) and Typhoon Pepeng (Parma), resulting in recovery and reconstruction requirements totaling US$4.4 billion, including US$2.4 billion in public spending needs (Government of the Philippines 2009). In the aftermath of the typhoons, the government of the Philippines with the World Bank and with support from GFDRR and partners (ADB, AusAID, JICA), undertook a post-disaster needs assessment (PDNA) with a series of recommendations to strengthen the country's resilience to natural disasters.

Approach: The World Bank's engagement has helped to strengthen the policy dialogue on DRM with the government of the Philippines. Based on these recommendations, the World Bank and GFDRR extended analytical support to formulate a disaster risk financing strategy to reduce the fiscal burden arising from the increasing costs of disasters, including the use of an innovative financing mechanism providing contingency financing in case of a national catastrophe. The PDNA was followed up with the development of a flood management master plan for metropolitan Manila to build the resilience of surrounding areas for future flood events, supported by GFDRR, AusAID, and JICA.

Action: In 2010 the Philippines signaled a policy shift from post-disaster response to prevention and mitigation. It enacted the DRRM Act and adopted a Strategic National Action Plan for Disaster Risk Reduction (SNAP) shortly thereafter, effectively institutionalizing a comprehensive and integrated approach to disaster risk reduction and management in the country. The law forms the base for the Disaster Risk Management Development Loan with a Catastrophe Deferred Drawdown Option (Cat-DDO), which fulfilled one of the key recommendations of the risk-financing study. This operation is the foundation for the Bank's ongoing policy dialogue on disaster risk reduction and management (DRRM) with the government and frames planned technical assistance programs. The operation, amounting to US$500 million to provide rapid liquidity in the event of a disaster, was signed by the Philippine government in September 2011. The full amount was disbursed in 2011 after Tropical Storm Sendong (Washi). The figure below illustrates various follow-up activities in recent years.

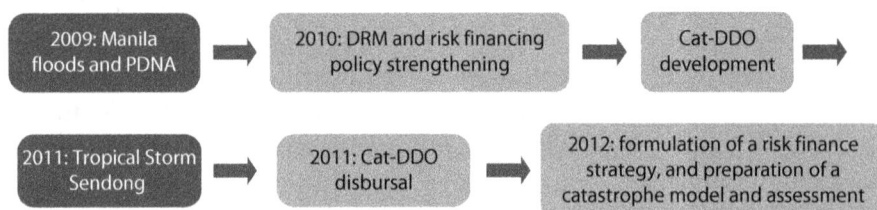

| 2009: Manila floods and PDNA | → | 2010: DRM and risk financing policy strengthening | → | Cat-DDO development | → |
| 2011: Tropical Storm Sendong | → | 2011: Cat-DDO disbursal | → | 2012: formulation of a risk finance strategy, and preparation of a catastrophe model and assessment | |

Next steps: The government is in the process of formulating its own risk finance strategy and has requested the World Bank's support in the preparation of a road map and work plan

box continues next page

Box 1.4 Strengthening the Philippines' Resilience to Disasters *(continued)*

on risk finance. In parallel, the World Bank with GFDRR funding is supporting the preparation of a catastrophe risk model and assessment to inform the design of a parametric risk finance instrument to offer financial protection for government's sovereign risks.

Key lessons learned:

- Do not wait for a disaster to happen: Reform the policy and action framework on DRM and invest in preparedness.
- Work together with partners to leverage investments and benefit from technical expertise.
- Develop a national strategy, focusing on cross-sectoral and cross-level coordination.
- Engage relevant stakeholders, such as the Ministry of Finance, on DRM to build fiscal resilience.
- Learn from past disasters and improve response.

Notes

1. Barthel and Neumayer argue that "[at a global scale] no significant trend is discernible. Similarly, we do not find a significant trend if we constrain our analysis to non-geophysical disasters in developed countries." See also Pielke et al. (2008).

2. There is a well-defined interdecadal variation in tropical cyclone activity in the northwest Pacific. For example, the period 1998–2010 was relatively inactive compared with 1989–98. It has been proposed that this is related to strong vertical wind shear and strong subtropical high pressure in the region of tropical cyclone genesis.

3. Source: GFDRR Disasters Portfolio Database (data as of June 30, 2012). The database includes all projects with any activity related to disasters, although it excludes many activities that have a non–disaster-related purpose but that may also help to reduce the impact of disaster.

4. Before 2006 the largest disaster projects always focused on reconstruction. Today there are also large ex ante projects: for example, the Catastrophe Deferred Drawdown Option for the Philippines (US$500 million, FY12), the India National Cyclone Risk Mitigation Project (US$255 million, FY10), the Jakarta Urgent Flood Mitigation Project (US$131 million, FY12), and the Istanbul Seismic Risk Mitigation and Emergency Preparedness Project (additional financing US$150 million, FY11).

5. The World Bank classifies the East Asia and the Pacific region into: Low-income economies: Cambodia, the Democratic People's Republic of Korea, Republic of the Union of Myanmar. Lower-middle-income economies: Fiji, Indonesia, Kiribati, the Lao People's Democratic Republic, the Marshall Islands, the Federated States of Micronesia, Mongolia, Papua New Guinea, the Philippines, Samoa, the Solomon Islands, Timor-Leste, Tonga, Vanuatu, Vietnam. Upper-middle-income economies: American Samoa, China, Malaysia, Palau, Thailand, Tuvalu. IDA: Cambodia, Kiribati, Lao PDR, the Marshall Islands, the Federated States of Micronesia, Republic of the

Union of Myanmar, Samoa, the Solomon Islands, Tonga, Tuvalu, Vanuatu. Blend: Mongolia, Papua New Guinea, Vietnam. IBRD: China, Fiji, Indonesia, Malaysia, Palau, the Philippines, Thailand.

6. Portfolio data as of August 28, 2012. Figure includes multidonor funds, donor-earmarked funds, and the Japan Social Development Trust Fund (SDTF), but excludes PHRD and AusAID IEAAG-funded projects.

7. HFA is the internationally accepted framework for disaster risk reduction and building resilience. This 10-year plan was adopted by 168 UN member nations in 2005 in Japan. The HFA provides a systematic approach to reduce vulnerabilities and identifies five Priorities for Action to reduce disaster risk: (1) making disaster risk reduction a policy priority and institutional strengthening; (2) risk assessment and early warning systems; (3) education, information, and public awareness; (4) reducing the underlying risk factors; and (5) preparedness for effective response.

References

ADB (Asian Development Bank). 2012. "Green Urbanization in Asia Key." Special chapter in *Key Indicators for Asia and the Pacific 2012*. Manila: Asian Development Bank.

Barthel, F., and E. Neumayer. 2011a. "A Trend Analysis of Normalized Insured Damage from Natural Disasters." *Climatic Change* 113: 215–37.

———. 2011b. "Normalizing Economic Loss from Natural Disasters: A Global Analysis." *Global Environmental Change* 21 (1): 13–24.

Chan, J. C. L. 2008. "Decadal Variations of Intense Typhoon Occurrence in the Western North Pacific." *Proceedings of the Royal Society A* 464: 249–72.

Climate and Development Knowledge Network. 2012. *Managing Climate Extremes and Disasters in Asia: Lessons from the IPCC SREX Report*. London. http://www.cdkn.org/srex.

Deltares. 2012. "Adaptive Delta Management Methodology." Presentation at the World Bank, Washington, DC, March.

Dilley, M., R. S. Chen, U. Deichmann, A. L. Lerner-Lam, and M. Arnold, with J. Agwe, P. Buys, O. Kjekstad, B. Lyon, and G. Yetman. 2005. *Natural Disaster Hotspots: A Global Risk Analysis*. Synthesis report, International Bank for Reconstruction and Development, World Bank, Washington, DC; Columbia University, New York.

EM-DAT. 2012. The International Disaster Database. Centre for Research on the Epidemiology of Disasters (CRED), Brussels. Accessed September 2012. http://www.emdat.be/database.

GFDRR (Global Facility for Disaster Reduction and Recovery). 2012. Internal Disasters Portfolio and Projects Database. Washington, DC. https://www.gfdrr.org/gfdrr/node/44.

Government of the Philippines. 2009. Philippines. Typhoons Ondoy and Pepeng: Post-Disaster Needs Assessment. Main Report. The Government of the Philippines: Manila. https://www.gfdrr.org/gfdrr/node/326.

Guha-Sapir, D., F. Vos, and R. Below, with S. Ponserre. 2012. *Annual Disaster Statistical Review 2011: The Numbers and Trends*. CRED, Brussels. http://cred.be/sites/default/files/2012.07.05.ADSR_2011.pdf.

Haldane, A. G., and V. Madouros. 2012. "The Dog and the Frisbee." Speech given at the Federal Reserve Bank of Kansas City's 36th Economic Policy Symposium, "The Changing Policy Landscape," Jackson Hole, WY, August 31. http://www.bankofeng-

land.co.uk/publications/Pages/news/2012/075.aspx; and http://www.bankofengland
.co.uk/publications/Documents/speeches/2012/speech596.pdf.

Hallegatte, S. 2009. "Strategies to Adapt to an Uncertain Climate Change." *Global Environmental Change* 19 (2): 240–47.

———. 2011. "How Economic Growth and Rational Decisions Can Make Disaster Losses Grow Faster than Wealth." Policy Research Working Paper 5617, World Bank, Washington, DC.

Little, R. G. 2009. "Managing the Risk of Cascading Failure in Complex Urban Infrastructures." In *Disrupted Cities: When Infrastructure Fails*, edited by S. Graham, 27–41. London: Routledge.

Perrow, C. 2011. "Fukushima and the Inevitability of Accidents." *Bulletin of the Atomic Scientists* 67: 44. http://www.yale.edu/sociology/faculty/pages/perrow/Fukushima11_1_11.pdf.

Pielke, R. A., Jr., J. Gratz, C. W. Landsea, D. Collins, M. A. Saunders, and R. Musulin. 2008. "Normalized Hurricane Damage in the United States: 1900–2005." *Natural Hazards Review* 9 (1): 29–42. http://sciencepolicy.colorado.edu/admin/publication_files/resource-2476-2008.02.pdf.

———. 2010a. *It Is Not Too Late: Preparing for Asia's Next Big Earthquake, with Emphasis on the Philippines, Indonesia, and China* [Policy Note], by P. I. Yanev. Washington, DC: World Bank.

———. 2010b. *Natural Hazards, Unnatural Disaster: The Economics of Effective Prevention.* Washington, DC: World Bank.

———. 2011. "East Asia Pacific, Disaster Risk Management Interview: Cascading Infrastructure Failures." Interview with Dr. Richard Little. http://web.worldbank.org/WBSITE/EXTERNAL/TOPICS/EXTURBANDEVELOPMENT/EXTDISMGMT/0,,contentMDK:22907219~menuPK:341053~pagePK:64020865~piPK:149114~theSitePK:341015,00.html.

———. 2012a. *Cities and Flooding. A Guide to Integrated Urban Flood Risk Management for the 21st Century.* Edited by A. Jha, R. Bloch, and J. Lamond. Washington, DC: World Bank.

. 2012b. *Climate Change, Disaster Risk, and the Urban Poor: Cities Building Resilience for a Changing World.* Edited by J. Baker. Washington, DC: World Bank.

———. 2012c. *Inclusive Green Growth: The Pathway to Sustainable Development.* Washington, DC.

———. 2012d. *Thai Flood 2011: Rapid Assessment for Resilient Recovery and Reconstruction Planning.* Washington, DC. http://www.gfdrr.org/gfdrr/sites/gfdrr.org/files/publication/Thai_Flood_2011_2.pdf.

———. 2012e. *The Great East Japan Earthquake: Learning From Megadisasters.* Knowledge Notes, Executive Summary, Washington, DC. http://wbi.worldbank.org/wbi/Data/wbi/wbicms/files/drupal-acquia/wbi/drm_exsum_english.pdf.

———. 2012f. *The Sendai Report. Managing Disaster Risks for a Resilient Future.* Washington, DC.

———. 2012g. *ASEAN: Advancing Disaster Risk Financing and Insurance in ASEAN Countries: Framework and Options for Implementation.* Vols. 1 and 2. Washington, DC: World Bank. http://www.gfdrr.org/gfdrr/sites/gfdrr.org/files/documents/DRFI_ASEAN_REPORT_June12.pdf.

CHAPTER 2

Strengthening Institutions and Outreach to Communities

Key Messages for Policy Makers

- Many countries in East Asia and the Pacific are grappling with challenges that limit their ability to effectively manage disasters, including inadequate institutional arrangements, poor coordination, insufficient capacity to manage risks, both ex ante and ex post, and restricted financial and human resources.
- Strengthening institutional coordination on disaster risk management (DRM) across sectors and stakeholders and capacity building across all levels of government is a priority for the region.
- To prevent duplication and dilution of institutional capacity among sectoral ministries and harmonize access to external funds, DRM and climate change institutions need to work in synergy.
- Building on existing community-based interventions and social protection systems provides an opportunity for countries to achieve significant outreach of DRM programs at the community and household levels and to reduce the socioeconomic impacts of disasters that disproportionately affect the most vulnerable segments of society.

Where Are We Now?

East Asia and the Pacific has the second largest number of fragile and conflict-stricken states and regions after Africa. Several countries in the region have high rates of poverty (the Lao People's Democratic Republic, Mongolia, Republic of the Union of Myanmar, Papua New Guinea, Tonga), struggle to overcome the legacy of conflict (Cambodia), and/or face ongoing risks of political insecurity (Republic of the Union of Myanmar, the Solomon Islands, Timor-Leste). In addition, some countries face geographical and/or political isolation and have small and vulnerable economies, and, for most part, their governance systems are dealing with

---

This chapter was written by Zoe Trohanis and Eiko Wataya with input from Patricia Fernandes and Rachel Cipryk.

difficult combinations of natural resource revenue wealth alongside weak systems and institutions. In addition to growing regional inequalities between low- and middle-income countries, there are growing gaps *within* middle-income countries in East Asia. Although some countries and subregions experience rapid growth and income increases, others lag behind, which does little to promote stability.

The impacts of disasters are often felt more acutely in fragile states. Fragile states have weak institutions and low capacity and find it difficult to grapple with both instability and disaster response concurrently. Numerous fragile areas have recently been affected by both conflict and disasters. This was the case of Tropical Storm Sendong, which hit Mindanao in the Philippines in 2011, the tsunami that struck Aceh, Indonesia, in 2004, and Cyclone Nargis, which struck the Republic of the Union of Myanmar in 2008 (box 2.1).

Disasters disproportionally affect the poorest segments of society, particularly women, children, the elderly, and people with disabilities. Poverty and vulnerability to disasters are closely linked. Living in hazardous areas, poor and vulnerable populations are more exposed to the impacts of disasters. Lacking access to basic services, such as clean water, sanitation, and health services, also makes the poor more vulnerable to the impacts of disasters.[1] Finally, they face constraints in their ability to respond and quickly recover their livelihoods, because of limited assets and financial resources, the absence of formal or informal social safety nets and social protection mechanisms, and lack of access to financial services and appropriate disaster finance products such as insurance or affordable loans. This highlights the need to strengthen social protection mechanisms to reduce the socioeconomic impacts on the most vulnerable segments of society. Linking these mechanisms to DRM increases the effectiveness of DRM programs in contributing to poverty alleviation and sustainable development.

Box 2.1 Impact of Cyclone Nargis in the Republic of the Union of Myanmar

Cyclone Nargis, which struck the Republic of the Union of Myanmar in 2008, was one of deadliest tropical cyclones, killing an estimated 140,000 people in a low-capacity country with recent history of conflict.[a] After Cyclone Nargis struck, the government of the Republic of the Union of Myanmar joined with the World Bank to lead other bilateral and nongovernmental development partners, the Association of Southeast Asian Nations (ASEAN), and the United Nations in carrying out the Post-Nargis Joint Assessment. The assessment identified various priority actions to implement disaster risk reduction programs, including a comprehensive multihazard risk assessment in the short and medium terms to guide the reconstruction process, as well as future development. Cyclone Nargis highlighted the need for the country to undertake a range of actions for reducing, mitigating, and managing disaster risks in the future to reduce hazard risks. The recovery effort provides an opportunity to strengthen existing or establishing new institutional, legislative, and financial arrangements for comprehensive DRM.

Source: Tripartite Core Group 2008.
a. Includes missing population; ASEAN (2009).

Linkages between DRM and social protection systems, which both build disaster resilience and respond effectively in post-disaster contexts, are limited in East Asia and the Pacific. The range of social protection coverage and sophistication in the region is wide, with middle- and high-income countries tending to have more comprehensive systems, including integrated social insurance and social assistance programming with emerging labor market programs, and lower-income countries focusing primarily on safety nets, relying often on donor support for their success. The 2009 food, fuel, and financial crises and the increasing cost of natural disasters have seen governments and planners return to well-funded, effective, and sustainable social protection programs. Although social protection programming is increasingly being linked with longer-term crises such as the 2009 crises, there is still unexplored potential to use these systems to respond to rapid-onset disasters.

Community-driven development (CDD) has been used successfully in East Asia and the Pacific to build community resilience and enable swift community reconstruction, but efforts could be significantly expanded. CDD emphasizes community control over planning decisions and investment resources,[2] it empowers local decision making, and it brings resources to communities efficiently (Wong 2012). Unlike traditional approaches, CDD enables communities and local institutions—rather than central governments—to take the lead in identifying and managing community-level investments. CDD approaches for smaller-scale disaster preparedness investments have proven to be consistently cost-effective. For example, in the Philippines, cost savings ranged from 8 percent for school buildings to 76 percent for water supply investments, when compared with traditionally implemented infrastructure (Araral and Holmemo 2007).

CDD has a long history in East Asia and the Pacific, with programs established as early as 1998 in Indonesia, in 2002 in the Philippines, and more recently in Lao PDR. Rapidly reaching national coverage, in Indonesia these programs were successfully used for post-conflict and post-disaster response following the 2004 tsunami. In Indonesia and the Philippines, CDD interventions include DRM elements. Lao PDR and the Philippines are currently scaling up CDD programs nationally. Interventions focusing on remote rural areas with significant numbers of ethnic minority populations are in place in China and Vietnam. The CDD program in the Solomon Islands is about to integrate disaster and climate risk management into community development.

Substantial progress has been achieved in developing national policy and legal frameworks for DRM in East Asia and the Pacific. In 2009 the World Bank carried out a series of assessments for Cambodia, Indonesia, Lao PDR, the Philippines, and Vietnam as well as a few Pacific island countries (the Marshall Islands, Papua New Guinea, the Solomon Islands, and Vanuatu) to provide a strategic overview of ongoing disaster risk reduction (DRR) initiatives in the region and identify gaps and opportunities for countries and the region as part of a comprehensive DRM program. A DRM law, act, or decree that establishes DRM agencies or committees is in place, for example, in Indonesia, Lao PDR, the Marshall Islands, Papua New Guinea, the Philippines, the Solomon Islands, and Vanuatu (see box 2.2 for examples).

Box 2.2 Examples of DRM Legislation in the Region

Indonesia: After the 2004 Indian Ocean tsunami, Indonesia enacted the Law on Disaster Management (Law 24/2007), which outlines the principles, responsibilities, organization, and implementation of the national DRM system, including the role of international organizations. The Strategic National Action Plan for Disaster Risk Reduction in Cambodia 2008–13 was launched in March 2009. Prepared by the National Committee for Disaster Management and Ministry of Planning, this strategy has been formulated to serve as a road map for development of institutions, mechanisms, and capacities of disaster management committees at all levels.

Philippines: The Disaster Risk Reduction and Management (DRRM) Act (2010) supersedes Presidential Decree No. 1566, which marks a shift in policy from post-disaster response to prevention and mitigation. The act also established the National Disaster Risk Reduction and Management Council as the coordinating body, with the Secretariat located in the Office of Civil Defense, with councils cascading from the regional to provincial to local councils. To complement the DRRM Act, the government of the Philippines has also formalized the implementing rules and regulations of the DRRM Act, the National Disaster Risk Reduction & Management Framework, and the NDRRM Action Plan, which translates the country's commitments to the Hyogo Framework for Action (HFA).

Solomon Islands: The National Disaster Risk Management Plan (2010) provides the institutional framework for coordination on all matters pertaining to DRM at national to local government levels, marking an effort to embrace a more holistic risk management approach.

Vanuatu: The nation is about to establish a National Advisory Board (NAB) for DRR and climate change, which will act as a central coordinating body aligning and harmonizing the DRR and climate change agendas in a more effective and efficient manner within the context of a capacity-constrained small island developing state. The Secretariat to the NAB will be equipped with project implementation capacity and enable it to act as the National Implementing Agency to channel future climate funding.

Vietnam: Approved by the government in November 2007, the National Strategy for Natural Disaster Prevention, Response, and Mitigation to 2020 marks the shift from post-disaster response to a more proactive approach focused on risk reduction. The strategy lays out Vietnam's primary DRM objectives, focusing largely on water-related disasters. DRM is integrated into Vietnam's Poverty Reduction Strategy and Country Development Plans.

Source: Based on World Bank/GFDRR 2011.

Some countries in East Asia and the Pacific suffer from weak institutional arrangements and poor coordination across sectors and different levels of government. Commitment to effective coordination varies across the region and can be limited by a range of constraints on capacity, funding, and political will. In Thailand the post-disaster needs assessment following the 2011 floods revealed coordination gaps between provincial and local administrative structures as well as with firms and associations in the private sector (World Bank 2012c): "Thailand's water sector is complex, with many agencies involved, but without significant coordination amongst them, or sufficient legislation to support the establishment of a single agency with oversight of the sector as a whole" (World Bank 2012c, 78). The government of Indonesia recently adopted a more comprehensive approach to coordination of disaster response covering ex ante and ex post stages of disasters through the National Disaster Management Agency (BNPB). It is also actively promoting the involvement of local governments and communities in DRM. Some countries are also at risk of diluting or duplicating their institutional capacities because of parallel arrangements for DRM and climate change adaptation at the national and local levels.

Limited progress has been made to provide adequate resources and support local-level implementation (UN 2009, 77; UNISDR n.d.). Given that local governments play a vital role as first responder in the event of a disaster and key stakeholder in post-disaster reconstruction and recovery, and in the implementation of preparedness and mitigation measures, it is important that they allocate sufficient and flexible resources and capacity-building efforts to empower them in these roles. Despite the immense progress made in many countries in East Asia and the Pacific to decentralize authority and resources to local governments, inadequate capacity and human resources to implement DRM efforts at the local level remain high. Evidence suggests that in countries in East Asia and the Pacific, limited funds are allocated for DRM (World Bank 2012a), although available data are limited. DRM expenditures are often difficult to track; most preventive measures are embedded in the design and construction of infrastructure or other sectoral spending (World Bank 2010, 106–10). Funding comes from a range of sources. In Lao PDR, funds are mobilized from the national and local budgets in the event of disasters, and the government earmarks a limited amount of the budget for emergency response per year. The government also does not have a national disaster relief reserve fund to provide funding for emergency response or recovery activities. In the case of Indonesia, the national budget allocation for DRM quadrupled from 2001 to 2007 in response to major disasters in Aceh and Java (see figure 2.1). It has since decreased, which suggests that most of the spending was for response and recovery. Looking at the sectoral budget allocations it is difficult to analyze the extent to which DRR is fully mainstreamed in regular development programs. A new government regulation on Funding and Management of Disaster Assistance stipulates three categories of funding: a contingency fund, an on-call budget, and social assistance funds. However, a comprehensive risk-financing system has not yet been put in place.

Where Do We Want to Be?

Enforce institutional and legislative frameworks that give clear guidance on risk management actions. Reform of the institutional and legislative arrangements for DRM to mainstream risk reduction and mitigation across all sectors and levels is essential for countries to become successful in addressing the underlying drivers of risk (UN 2011). This is particularly important for low-capacity, fragile, and post-conflict countries. National strategies should be complemented by concrete

Figure 2.1 Post-Disaster and Pre-Disaster Spending Levels

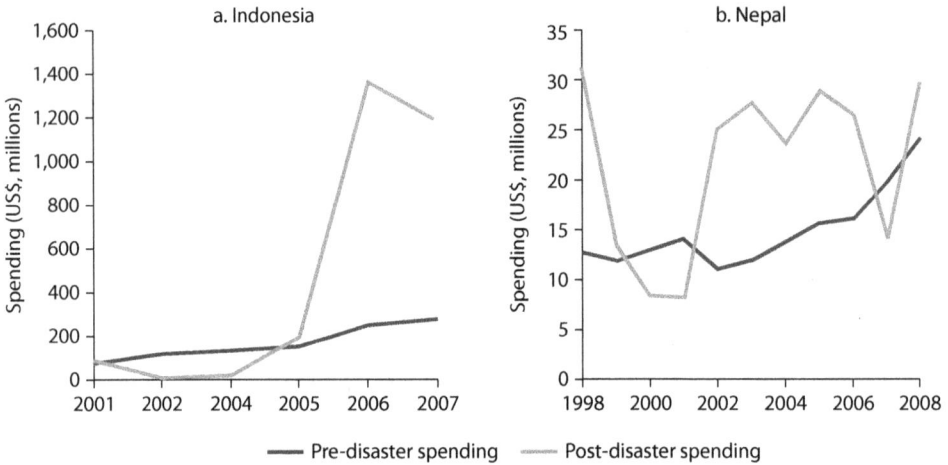

Source: de la Fuente 2009 cited in World Bank 2010, 107.

Box 2.3 Lincolnshire Mapping of Critical Assets Case Study

During 2010, Lincolnshire's Critical Infrastructure and Essential Services Group held a series of workshops looking at critical infrastructure along its coastal strip. Local representatives and asset owners, including Anglian Water, British Telecom, CE Electric, and five of the local drainage boards, attended these workshops. The results will feed into the local Multi-Agency Flood Plan community impact assessments.

During the workshops, organizations were asked to look at four issues: identifying assets, assessing their ability to continue to provide services during a flood, highlighting interdependencies between asset owners, and service restoration time frames. The workshops were an opportunity to review and update Lincolnshire's geographic information system, which already contains sites including telephone exchanges, electricity substations, and water and waste assets, together with vulnerable community assets such as blue light services, rest centers, and schools. Key locations were highlighted in which the impact of community flooding would be significantly worsened by infrastructure failure.

Source: UK Cabinet Office 2011.

---

action plans, reaching from the national to the community level. Laws and regulations should be accompanied by enforcement mechanisms and positive incentives. National response plans often include procedures on how to request international assistance, while local plans provide greater detail on evacuation and shelter plans. These plans should be developed through a process of stakeholder consultations (box 2.3) to ensure effective coordination during and in the aftermath of a disaster.

Establish adequate funding, coordination, accountability, and stakeholder participation arrangements. Reliable and sufficient funding is needed for preventive action and post-disaster recovery. It is important for governments to develop reliable, multiyear DRM-financing strategies as part of their development planning. Better coordination mechanisms strengthen implementation across the range of stakeholders. Monitoring and evaluation systems allow responsible agencies and institutions to analyze ex ante and ex post efforts and their impacts on the ground in relation to the needs of the population and adjust them accordingly. This is also important for measuring progress toward an effective DRM. Giving local and affected populations a voice in DRM through community-based DRM approaches has proved successful in engaging communities to encourage self-help efforts of building resilience at the local level in a comprehensive manner.

Promote a holistic approach to disaster management. To effectively support the most vulnerable populations, it is important to link social protection and CDD programs with DRM. This can help build climate and disaster resilience at the local level, in collaboration with vulnerable communities. CDD and social protection programs can be also very effective vehicles for channeling post-disaster support because they rely on existing functioning systems that can quickly disburse resources to known vulnerable households and communities. Additionally, they link the official DRM program with the social programs' large network of case workers or facilitators that can collect and monitor information on local-level needs and gaps. Existing experiences in CDD and DRM in the region and globally underscore the importance of having systems and networks in place to rapidly scale up operations in the aftermath of disasters (box 2.4). In Madagascar, the Social Development Fund (FID) has been used for both community-based disaster reconstruction as well as addressing the issues of food

---

Box 2.4 Indonesia: Using CDD Programs to Respond to Disasters

The National Program for Community Empowerment Mandiri (PNPM Mandiri) is the Indonesian government's flagship community-based poverty alleviation program. PNPM Mandiri uses a CDD approach, providing direct block grants of about US$20,000, financing small-scale socioeconomic infrastructure, education and health activities, and microloans for women's savings groups. Following the Asian tsunami in 2004, PNPM developed a comprehensive set of operational procedures to expedite and support disaster recovery, which are essentially modifications to the program's existing operations manual, speeding up planning and expanding the menu of possible activities to be implemented with community grants to account for special needs in emergency situations. The PNPM program also has a component called "contingency for disaster risk response." Key changes to the normal PNPM cycle and operation included the following:

Shortening the village consultation and assessment period: Before the tsunami, the time allocated to village consultations in the project cycle was four to five months. To respond to community needs after the tsunami, the cycle was shortened to one to two months. The team used a rapid "damages and loss assessment" instead of the typical Participatory Social Analysis approach to identify potential projects.

box continues next page

Strong, Safe, and Resilient • http://dx.doi.org/10.1596/978-0-8213-9805-0

Box 2.4 Indonesia: Using CDD Programs to Respond to Disasters *(continued)*

Strengthening coordination and administration systems to deal with the scaling up of activities: (1) Hiring of additional trained staff, in particular facilitators, consultants, and other support staff, to supplement people already on the ground. To handle the disaster situation, new types of staff were required, such as computer operators for data management and "information facilitators" responsible for data collection, information sharing and dissemination, and communication with stakeholders and external partners; (2) coordinated interventions with nongovernmental organizations active in reconstruction activities and to prevent overlap; and (3) streamlining procurement and disbursement procedures following rapid response procedures. The PNPM experience also highlights the importance of retaining key staff and budgeting for local wage increases. International agencies that are generally planning a short- to medium-term presence tend to drive up local wages because they are prepared to pay significantly higher local wage rates than those before the tsunami to secure key personnel quickly. In Aceh, salaries almost doubled for certain clerical and administrative staff during the first 10 months after the disaster. Wages for construction workers also doubled.

Increasing grant ceilings and channeling additional resources to affected villages using existing systems. Given the special needs of post-disaster areas (Aceh and Nias Islands), additional grants of up to Rp 7 billion (US$525,000) were provided to affected areas. Almost all of the villages affected by the tsunami had funds remaining in their communal accounts that had yet to be disbursed. These villages were permitted to allocate 25 percent of those remaining funds to any pressing social needs they deemed urgent and necessary. Eligible items to be purchased were detailed in "procurement packets," and then funds were distributed to those in need. In addition to the first allocation of Social Funds, the affected villages were also permitted to allocate another 25 percent of the next cycle of PNPM funding to their Social Fund if they decided there were still families and individuals in need of assistance. New villages joining PNPM or those villages without remaining funds were also entitled to allocate 25 percent of their block grants for social purposes as long as they had been affected by the tsunami. Labor-intensive projects were also prioritized.

Source: Earth Systems Laos 2012.

security. DRM principles also have been integrated into the Rural Development Support Project (PSDR), a community-driven development program complementary to FID supporting agriculture, national resource management, and rural development.

What Needs to Be Done?

Although there are common opportunities for the countries in East Asia and the Pacific to strengthen their risk management, capacity, and scope of activities, there is no one-size-fits-all solution. The recommendations below should be adjusted to individual country contexts with respect to institutional structure, hazard, exposure and vulnerability profiles, and socioeconomic situation.

Although many countries would benefit from strengthening each of these areas, those with particularly weak governance, history of fragility, and/or low capacity might focus first on creating a framework for DRM and build the rest as stability is established and capacity is strengthened.

1. **Understand institutional arrangements and needs.** An emerging priority, especially for countries with a history of fragility and/or low capacity, such as Lao PDR, Republic of the Union of Myanmar, Timor-Leste, and some of the Pacific island countries, as well as for countries with complex institutional arrangements, is to identify the complete set of stakeholders involved in DRM and to ascertain whether particular agencies or departments need strengthening or whether new arrangements are needed. It is important to understand that mandates, responsibilities, capacities, resources, effectiveness of relevant institutions, and coordination mechanisms should be assessed, including factors that can increase or decrease their influence, such as changes in political administrations, new data available, or regional networks. Mongolia is currently in the process of assessing its institutional setup (World Bank/GFDRR 2011, forthcoming). Findings highlight the necessity to (1) increase capacity for application and implementation of the existing policy framework for DRR, (2) strengthen monitoring and analysis of hazard risks and dissemination structures such as early warning systems, and (3) improve coordination across government agencies and with donors to enable an assessment of Mongolia's preparedness plans and decision-making capacities. Identifying specific gaps and needs is useful for aligning resource plans and programs between government, development partners, and international financial institutions.

2. **Provide adequate and reliable funding for DRM.** To address gaps in funding that influence ex ante preparedness and post-disaster response measures, planners should start with a stocktaking exercise to review current financing mechanisms, past expenditures and budget gaps, budget arrangements, and DRM allocations. Tracking expenditures at subnational levels is particularly important to be able to identify the needs of implementing agencies. Investing in cost-benefit analysis of prevention and mitigation activities is also a good practice to target investments based on a quantitative analysis.

3. **Harness community knowledge and power.** Partnerships and collaboration with nongovernmental and civil society organizations should be harnessed to a greater extent. Local academic institutions such as universities also become beneficial technical resources that can be tapped. Development of partnerships with these groups can assist in adapting international DRM practices to the local context. South-South learning activities can also be instrumental in this regard.

4. **Focus on vulnerable communities and households through mainstreaming DRM into CDD and social protection interventions.** As the region scales up

CDD programs, a critical window of opportunity opens to establish systems that can build resilience and can be easily scaled up when a disaster hits and quickly scaled down after urgent post-disaster needs are met. Linking with these systems also taps into a large database of information on poverty trends, such as areas with large concentrations of poor, that are invaluable in targeting both resilience-building interventions and post-disaster relief and reconstruction funding.

5. **Use learning and knowledge to strengthen stakeholders' capacities.** To strengthen the institutional capacity in the region, continue to cultivate a learning culture among government officials, policy makers, and practitioners and apply new knowledge in local contexts. To achieve this, mobilize financial and human resources to provide necessary services, demonstrate strong leadership, and strengthen implementation on the ground. Talent and knowledge retention within institutions poses a serious challenge for many countries in East Asia and the Pacific. To reverse this trend, it is crucial to assist officials and practitioners to enhance internal knowledge and technical transfer while minimizing knowledge and skill loss. A sense of ownership and commitment is a key step toward sustainable institutional capacity building. Supporting leaders and role models can be an effective approach to enhance the transferability of practical skills and experiences.

6. **Set up benchmarks and monitoring and evaluating systems.** Monitoring and evaluation should be legally anchored and integrated within national development-planning processes to ensure ownership and sustainability for the process and the system. In post-disaster and post-conflict situations, a multihazard approach to DRR should be factored into policies, planning, and programming related to sustainable development, relief, rehabilitation, and recovery activities (UNISDR n.d., 26). Partnerships with actors at different levels are important to strengthen institutions and capacities and, in the long term, promote sustainable institutions. The HFA offers a set of indicators to benchmark progress achieved in terms of institutional capacity building. Areas of focus include operational, national, and multisectoral platforms for DRM; institutional capacities, systems, policies, and legislation for DRR; technical and institutional capacities for disaster preparedness; contingency plans; and identification of capacity and resource gaps. Local adaptation is needed to fit the range of factors, including development priorities, income levels, and hazard profiles.

How Can the World Bank Help?

Partnerships strengthen cooperation and knowledge transfer, leverage funds, and increase aid effectiveness. The World Bank promotes partnerships with international organizations, national agencies, donors, cities, communities, academia, and learning institutions to strengthen cooperation and regional DRM partnerships and help to raise risk awareness, and encourage the development of joint programs and initiatives. Regional cooperation facilitates access to cutting-edge technical knowledge, capacity-building efforts for disaster prevention, mitigation, and response, and dissemination and knowledge transfer on DRR, for example, through South-South knowledge sharing and convening platforms for further actions. For example, a regional partnership with ASEAN, UNISDR, and the World Bank in 2009 supported country-level and regional investments in ex ante DRR through an enabling legislative, regulatory, and financing framework (box 2.5). The Bank also collaborates with the Asian Development Bank (ADB), Asian Disaster Preparedness Center (ADPC), Asian Disaster Reduction Center

---

Box 2.5 Partnership with ASEAN

The Association of Southeast Asian Nations (ASEAN) Secretariat, the United Nations International Strategy for Disaster Reduction (UNISDR), and the World Bank cooperation program builds on the experience of multidonor partnership support following the Indian Ocean earthquake and tsunami in 2004, the Yogyakarta earthquake in 2006, and Cyclone Nargis in the Republic of the Union of Myanmar in 2008. Continued collaboration after these events helped identify further assistance needs to support affected countries and to strengthen resiliency toward hazard risks in the region. Three parties agreed to set up a joint consultative mechanism at the working level to strengthen coordination in disaster relief, culminating in the Memorandum of Cooperation (MoC) in 2009.

As part of the MoC, technical support is provided to the ASEAN Secretariat on the implementation of DRR components of the ASEAN Regional Programme on Disaster Management (ARPDM) and the ASEAN Agreement on Disaster Management and Emergency Response (AADMER) along the lines of the HFA. Recent key activities include the ASEAN Disaster Risk Financing and Insurance Forum, and a series of Disaster Risk Management in East Asia and the Pacific: Distance Learning Seminar Series. The Global Facility for Disaster Reduction and Recovery (GFDRR) Track I service line (Global and Regional Partnerships), jointly run by the World Bank and the UNISDR, supports ASEAN in delivering the DRR component of the AADMER and ARPDM and special activities, including regional HFA progress reporting.

The ASEAN Disaster Risk Financing and Insurance Forum was held in November 2011 as a joint initiative of the ASEAN Secretariat, World Bank/GFDRR, and UNISDR. Supported by the government of Indonesia, the forum brought together over 100 senior-level policy makers from all 10 member countries, working in multiple sectors including disaster management, insurance, and finance to ensure a comprehensive discussion and initiate the development of a risk-financing road map for ASEAN as an effective means to manage the worsening financial impacts of disasters on member states. The forum served as a mechanism to strengthen the capacity of participants by sharing international knowledge and experience on catastrophe risk modeling and financing schemes. At the conference, ASEAN member states drafted a road map for capacity building and dialogue to develop a regional strategy for disaster risk-financing and insurance (World Bank 2012a).

(ADRC), Australian Agency for International Development (AusAID), Australia-Indonesia Facility for Disaster Reduction (AIFDR), Japan International Cooperation Agency (JICA), government of the Republic of Korea, Nanyang Technological University's Institute of Catastrophe Risk Management, Applied Geo-Science and Technology Division of the Secretariat of the Pacific Community (SOPAC), World Meteorological Organization (WMO), and others.

The World Bank provides knowledge and learning programs to build capacity of clients to be able to better plan, finance, and manage their DRM efforts at different levels. In the Philippines, the Bank-supported capacity-building program targeting the integration of DRM and climate change adaptation into one plan at the local level has been particularly successful: The Department of the Interior and Local Government is scaling up the program using its own funds with a budget programmed for 2013–15. In Lao PDR, after Tropical Storm Ketsana, in addition to assessment and preparation of the financial plan for recovery and reconstruction, training for government and international agencies was conducted to teach government staff how to carry out such assessments themselves in the future. After Typhoon Haima in 2011, the World Bank assisted the government in identifying capacity gaps and strengthening central and local governments to conduct post-disaster needs assessment (PDNA) and mitigate future flood risks. Coordination in the area of capacity building is necessary to synergize programs with other partner organizations effectively to deliver targeted training courses and avoid duplication of efforts. GFDRR is supporting these efforts.

The World Bank provides technical knowledge and timely analysis of pressing issues in the region related to DRM. With the world experiencing the fastest rates of urbanization in history, focusing on disaster risks in cities is key to safeguarding people and assets at risk. *Climate Risks and Adaptation in Asian Coastal Megacities*, jointly prepared by the World Bank, the Japan International Cooperation Agency, and the Asian Development Bank, strengthens the understanding of climate-related risks and impacts in coastal megacities in developing countries. The study considers Bangkok, Ho Chi Minh City, and Manila and their different environmental and socioeconomic characteristics to address the scale of impact and vulnerability, damage costs, and adaptation options, including climate change impacts, as well as in other coastal cities. In the Philippines, the Bank also supports a program to reduce vulnerability to flooding in metropolitan Manila, including preparation of the flood management master plan for the greater metropolitan Manila area. The program builds upon recommendations from the PDNA conducted after Tropical Storm Ondoy and Typhoon Pepeng in 2009, which identified the need to formulate the master plan to optimize and prioritize institutional arrangements for flood management.

The World Bank promotes innovative approaches to DRM, giving communities a voice and linking DRM with community-driven and social protection programs. Following the 2004 Indian Ocean earthquake and tsunami, in Aceh and Nias, Indonesia, a community-based disaster risk management

---

Box 2.6 Using Social Protection Mechanisms to Respond to Disasters

Social protection programs can quickly identify poor and vulnerable households that need additional assistance after a shock and provide them with resources to reduce the gap between shock and response delivery. Households can be existing safety net beneficiaries or new beneficiaries identified by the existing social system and networks.

In Pakistan, the Citizen's Damage Compensation Program was designed to provide recovery compensation in the wake of the devastating 2010 floods. The program uses a national registry system in which all citizens and their wealth ranking are registered. For post-recovery relief the poorest and most vulnerable populations in flood-affected areas were given priority to receive compensation transfers. The benefits of this program include (1) using the national poverty database to identify which affected households were poorest and most in need and prioritizing them for targeting of scarce resources and (2) using an automated database for quick registration of the program and immediate disbursement of benefits through an electronic cash card.

In Ethiopia, the Productive Safety Net Program provides regular assistance to more than 8 million chronically food-insecure people as a condition for performing public works. The program is also designed to quickly scale up to provide additional assistance to existing beneficiaries and short-term assistance to nonbeneficiaries when a shock hits. Once the crisis has passed, the program can be scaled back down to its normal size. The disaster response component is also linked to a rainfall index that allows the government to trigger early response for slow-onset disasters, such as drought. The benefits of this arrangement include using an existing, functioning, and effective infrastructure for (1) targeting beneficiaries, based on poverty *and* shock impact, (2) rapid payment of benefits, (3) rapid and easy flow of resources from the federal level down to the community level, (4) providing that the work done in exchange for transfers contributes to community resilience, and (5) identifying the need for and triggering response in a timely manner using a preagreed-to index threshold.

---

approach to reconstruction has proven successful to rebuild infrastructure, homes, and the social fabric of affected communities. As a key partner in many CDD programs in the region, the World Bank can also help integrate DRM principles into the operational procedures in Lao PDR, the Philippines, the Solomon Islands, and Vietnam. Linking DRM and social protection systems has been successful in World Bank–supported projects in Ethiopia and Pakistan (box 2.6). Having proved to be efficient in their use of government systems across sectoral boundaries for post-disaster relief, these two examples highlight that contingency planning and additional financing can leverage existing networks of community institutions, facilitators, and channels to disburse funds. Chapter 7 provides further details about emergency contingency measures.

Strong, Safe, and Resilient • http://dx.doi.org/10.1596/978-0-8213-9805-0

Notes

1. Arnold in World Bank (2000); World Bank (2012b).
2. The World Bank currently supports approximately 400 CDD projects in 94 countries valued at almost US$30 billion. The largest number of projects is in Africa, followed by South Asia and Latin America. However, commitment amounts are highest in East Asia and the Pacific, Africa, and South Asia. Over the past 10 years, CDD investments have represented between 5 and 10 percent of the overall World Bank lending portfolio.

References

Araral, E., and C. Holmemo. 2007. "Measuring the Costs and Benefits of CDD: The KALAHI-CIDSS Project Philippines." World Bank Social Development Paper 102, World Bank, Washington, DC. http://siteresources.worldbank.org/EXTSOCIAL DEVELOPMENT/Resources/244362-1164107274725/3182370-1164201144397/3187094-1173195121091/SDP-102-Jan-2007.pdf.

ASEAN. 2009. *Fourth ASEAN State of the Environment Report*. Jakarta.

Earth Systems Laos. 2012. "Options Paper on Gender-Sensitive DRM in the Philippines." World Bank, Washington, DC.

Tripartite Core Group. 2008. *Post-Nargis Joint Assessment*. http://gfdrr.org/docs/Post-Nargis_Joint_Assessment_July_21__08.pdf.

UN (United Nations). 2009. *United Nations Global Assessment Report on Disaster Risk Reduction*. New York: United Nations.

———. 2011. *United Nations Global Assessment Report on Disaster Risk Reduction*. New York: United Nations.

United Kingdom Cabinet Office. 2011. *Keeping the Country Running: Natural Hazards and Infrastructure: A Guide to Improving the Resilience of Critical Infrastructure and Essential Services*. London: United Kingdom Cabinet Office.

United Nations International Strategy for Disaster Reduction (UNISDR). n.d. *A Practical Guide to National HFA Monitoring and Review through a Multi-Stakeholder Engagement Process 2011–2013*. New York: United Nations.

Wong, S. 2012. *What Have Been the Impacts of World Bank Community-Driven Development Programs?* CDD Impact Evaluation Review and Operational and Research Implications, World Bank, Washington, DC. http://microdata.worldbank.org/index.php/catalog/1047/download/20976.

World Bank. 2000. *Managing Disaster Risk in Emerging Economies*. Edited by A. Kreimer and M. Arnold. Washington, DC: World Bank.

———. 2010. *Natural Hazards, Unnatural Disaster: The Economics of Effective Prevention*. Washington, DC: World Bank.

———. 2012a. *ASEAN: Advancing Disaster Risk Financing and Insurance in ASEAN Countries: Framework and Options for Implementation*. Vols. 1 and 2. Washington, DC: World Bank. http://www.gfdrr.org/gfdrr/sites/gfdrr.org/files/documents/DRFI_ASEAN_REPORT_June12.pdf.

———. 2012b. *Climate Change, Disaster Risk, and the Urban Poor: Cities Building Resilience for a Changing World*. Edited by J. Baker. Washington, DC: World Bank.

————. 2012c. *Thai Flood 2011: Rapid Assessment for Resilient Recovery and Reconstruction Planning*. Thailand: World Bank.

World Bank, Global Facility for Disaster Reduction and Recovery (GFDRR). 2011. *Disaster Risk Management Programs for Priority Countries*. Washington, DC: World Bank.

————. Forthcoming. *Mongolia Country Profile*. World Bank, Washington, DC.

CHAPTER 3

Risk Identification

Key Messages for Policy Makers

- Risk identification is the foundation for sustainable disaster risk management (DRM). It informs the prioritization and design of risk reduction investments, emergency preparedness, financial protection measures, and sustainable recovery and reconstruction. The process of assessing risk can leverage and add value to other key DRM activities such as hydrometeorological (hydromet) services and risk-financing mechanisms.
- Robust risk assessments rely on an investment in data sets such as past disaster impacts, estimations of future disaster events, baseline exposure and critical infrastructure, and demographic and socioeconomic information. High-resolution data should be collected and validated to inform development planning and DRM programs at the local level.
- In-country expertise is needed to develop and then continually improve and update risk information. Capacity should be built across all levels of government and in technical institutions to contribute to producing, sharing, and using risk information.
- Sharing and effectively communicating risk information enables a broad range of stakeholders in government, the private sector, and civil society to make better decisions to manage risk. In many cases technology is not the limiting factor; rather, institutional relationships and the political will to open information to the public need to be fostered.

Where Are We Now?

In East Asia and the Pacific, there is a critical need to invest in robust risk identification methodologies and tools. Identification of hazard, exposure, and vulnerability lies at the core of a holistic approach to DRM. Risk identification, conducted through risk assessments at the national, regional, city, and community levels, informs prioritization and design of structural and nonstructural risk

This chapter was written by Abigail Baca and Liana Zanarisoa Razafindrazay, with input from Michael Bonte-Grapentin and Anna Burzykowska.

reduction investments, emergency preparedness, financial protection measures, and sustainable recovery and reconstruction (figure 3.1). Different types of risk identification approaches can be selected based on the desired risk reduction tools listed in Step 5, as well as the constraints imposed by the availability of resources. In the case of scarce data and modest resources to collect new information, many projects rely on relatively low-resolution risk information that has been created at the international scale. These data can be used in education and capacity building to raise awareness about the general disaster risk trends and as a first step in prioritizing further risk assessment work. However, such low-resolution risk results cannot be used for some other DRM activities. Risk reduction tools such as risk-sensitive spatial planning, resilient design of critical infrastructure, or fiscal risk analysis for disaster risk-financing require a more detailed level of analysis.

As part of the risk assessment process, tools and data are needed to quantify the drivers of risk. This involves a spatial analysis of potential **hazards,** estimation of geographic extent, severity, frequency, and its potential impacts through and understanding of **exposure,** and the **vulnerability** of those elements. These key data are often developed as geospatial information, which allows the risk analysis results to be mapped and more easily communicated. Examples a and b in figure 3.2 display seismic hazard and critical infrastructure asset exposure maps. Example C shows the risk results from the analysis of the hazard and exposure combined with vulnerability or damage functions to yield the distribution of average annual loss.

Risk information has many uses both in DRM and beyond. The maps and data produced through the risk identification process can be leveraged for many purposes. The exposure data, including maps of population, buildings, and crops, are valuable as stand-alone tools for land-use planning and asset management.

Figure 3.1 Elements of Risk Identification and Risk Reduction in DRM

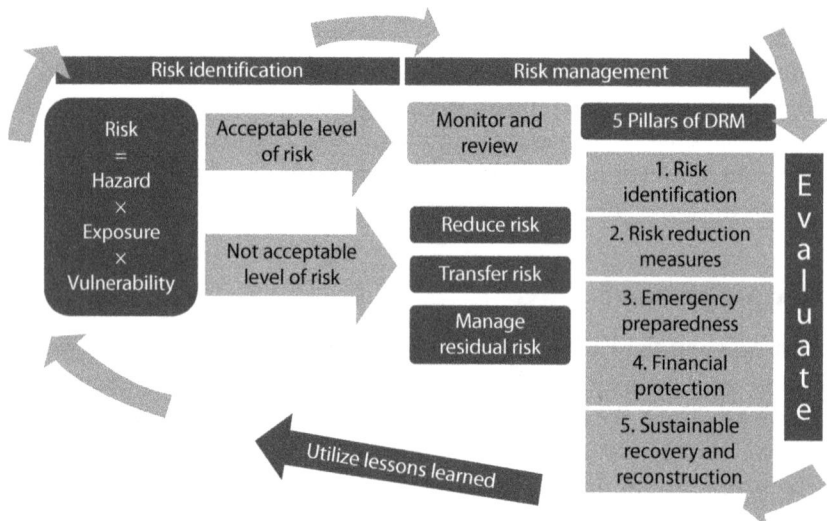

Source: Authors.

Figure 3.2 Hazard, Exposure, and Risk Maps for Papua New Guinea

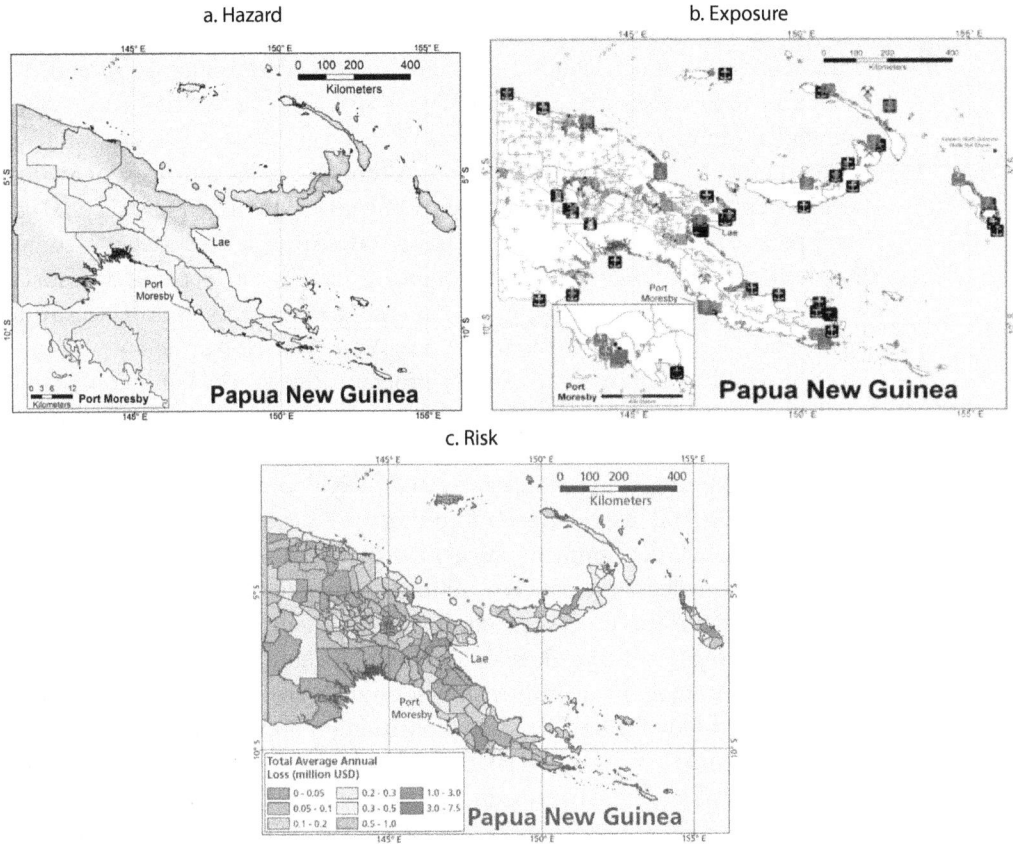

a. Hazard

b. Exposure

c. Risk

Source: World Bank 2011.

Hazard maps, such as the one in figure 3.2, panel a, offer valuable information to be integrated in building codes and construction guidelines. Scenario-based hazard information can be used for multiple purposes, from catastrophe risk financing, through emergency preparedness planning and training, to the emergency management information systems they feed.

Countries in East Asia and the Pacific face different needs and constraints in the area of risk identification. Many countries lack the financial and technical resources to systematically develop robust risk information, and they face institutional constraints that make it difficult to share, manage, and analyze risk information. These challenges make it difficult for countries in East Asia and the Pacific to effectively use risk information in DRM activities, and more generally in development planning. The Hyogo Framework for Action (HFA) offers a set of indicators for monitoring the progress in risk identification efforts (UNISDR 2009, 2010).[1] The following overview of the key challenges in this section is based on the results of the survey of countries reporting on their progress in this area (UNISDR 2011c, see also UNISDR 2011b).[2]

Lack of knowledge and technical capacity limits the ability of many countries in East Asia and the Pacific to assess risks accurately. Risk assessments cannot be regarded as a one-time investment. Effective risk assessments inform risk identification by integrating scientific modeling with localized knowledge and field validation. In East Asia and the Pacific, fewer than half of all countries have access to national multihazard risk assessments or invest in DRM-related research at a national budget level (UNISDR 2011a). Although significant investments have been made in most middle-income countries, such as Indonesia and the Philippines, less-developed and small island states struggle to develop in-country technical capacity. Higher capacity countries have been focusing on capacity building at a subnational level, involving local governments, research institutes, universities, and civil society organizations. As part of an Australian Agency for International Development (AusAID) program, the World Bank conducted a needs and capacity gaps analysis for Jakarta, Indonesia, and Can Tho, Vietnam, as part of highlighting challenges to risk identification and risk-based decision making in the urban context (World Bank 2012). Key findings indicate that enabling those efforts are still marginal, and more support is needed from government, the private sector, and development partners.

Lack of robust baseline data is a common constraint in countries in East Asia and the Pacific in the process of improving the quality of risk information for decision making. Capturing detailed information about the impacts of historical events is important. Frequently risk assessments are conducted post-disaster, mainly using funding from disaster reconstruction and recovery. Consequently the findings tend to be closely related to the most recent disaster, failing to cover the full range of potential hazards in the country. Based on the HFA survey, only 40 percent of responding countries in East Asia and the Pacific possess a historical event loss catalog or database. Although large-scale disasters trigger reconstruction and recovery attention from the national government and donors, smaller scale, more recurrent events can cumulatively have a significant impact on both local and national resources. Accounting for only the large events yields incomplete historical disaster information, which can bias key decisions, making it difficult to achieve sustainable results in disaster risk reduction.

Sharing the results of risk identification between government agencies and with a wider range of stakeholders is not common practice in most countries in East Asia and the Pacific. With the rapid development of technology, many countries in East Asia and the Pacific are launching online data portals to support information sharing. However, to date, only 40 percent of reporting countries have publicly available risk information (see CRED and USAID 2012 or Weber 2011). Barriers to information sharing include capacity and infrastructure gaps as well as political resistance to openness and security concerns. These challenges can be particularly great in the least-developed, post-conflict, or fragile states. In middle-income countries, coordination may be lacking between line ministries, including the procurement of new data, which implies duplication and unnecessary waste of resources. Sharing risk information in a transparent way can help to build a culture of accountability.

Where Do We Want to Be?

Integrate risk identification into the decision-making process. Governments need to know whether their investments will bring the benefits they aspire to, including if they decrease or, on the contrary, create new risks. The private sector needs accurate information about hazards to develop financing risk management strategies, including the evaluating and necessity of insurance coverage and informing land valuation. Industries need to know whether their prospective plants are located in a cyclone-prone zone, which could require conforming to specific land-use or building standards. Communities need to know whether their homes, businesses, and schools are located in dangerous zones. More importantly, at-risk populations need to know evacuation routes and destinations during a disaster event. All those questions can be answered through risk identification. Integrating risk information in the decision-making process can provide vital information about available options and, through informed action, can bring a higher return on structural and nonstructural investments. With the rapidly changing urban environments of East Asia and the Pacific and uncertainty linked to climate change impacts, there is an urgent need for a dynamic and flexible approach according to which development interventions and investments are identified, prioritized, and designed based on an understanding of changing exposure and vulnerabilities and consideration of hazard uncertainty (figure 3.3).

Strengthen institutional frameworks and coordination across levels and sectors to develop and use risk information. The absence of well-defined roles and tasks for governmental institutions is a barrier to effectively responding to and

Figure 3.3 Dynamic Decision-Making Process

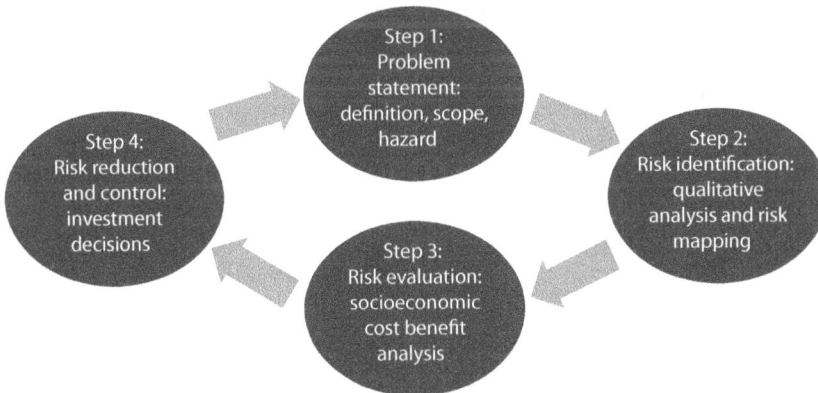

Dynamic analysis considers
- Current conditions
- Future variations in climate and urban exposure
- Multiple investment options

Dynamic analysis
- Prevents lock-ins in large investment projects
- Allows for feedback to improve risk identification

Source: Adapted from World Bank 2012.

preparing for disasters. Defining the roles of institutions is crucial, not only during the emergency response, but also for implementing preventive measures, including the risk identification process itself. Traditionally, local institutions are the closest partner for communities and the first responder in case of a disaster. It is important to furnish them with the adequate institutional, financial, and technological tools to fulfill their responsibilities. As described in the previous chapter, national governments need frameworks that clearly assign roles horizontally—between the different governmental departments and sectors, as well as vertically—from the national agencies to the local institutions and communities. All levels of institutions have a role to play in risk identification and DRM in general, recognizing the particular need to directly engage and work with the communities (see figure 3.4).

Develop baseline data at the local level with participatory mapping tools and methods. Baseline data, such as past hazard events, local-level administrative unit names and boundaries, demographic structure, topography, and land-use land cover and location of critical infrastructure, are crucial because they give a context to support decision making at the national, regional, and local levels for urban development, land-use management, emergency response, and other activities. Lack of high-resolution baseline data is a common problem in East Asia and the Pacific because most quantitative risk assessments are conducted at a coarse, national scale. Participatory mapping is an affordable method to collect information at the community level. Grassroots engagement is an effective way to raise awareness among populations, enabling communities to directly contribute to the production and dissemination of risk information. Participants are trained in the use of collaborative mapping technology, and local knowledge is integrated into the risk identification process. Partnerships and interactions between communities, local authorities, and government agencies improve the effectiveness of communication during and after a disaster. In Indonesia, the National Disaster Management Agency with partners organized in 2011 a province-wide participatory mapping exercise as part of contingency planning in Jakarta (box 3.1).

Figure 3.4 Risk Identification across Different Levels

National government: Involves government agencies across different sectors, including technical institutions that provide hydromet and geological services, and is coordinated by one central DRM agency, which distributes roles and resources to regional and local authorities for preparedness and response.

Local government: Regions and provinces assess risks at scales that go beyond subnational boundaries to coordinate and liaise with the central government for resources and with local authorities for preparedness and response. Meanwhile, local or district authorities assess localized hazards and exposure and engage communities in their activities.

Communities: Communities contribute baseline information about their own environment (past hazards, critical infrastructure, and individual assets) and societal vulnerability (poverty, gender, education, livelihoods) characteristics through community mapping activities and community participation, in coordination with local authorities.

Source: World Bank staff.

Box 3.1 Creating Critical Infrastructure Baseline Data with Participatory Mapping

As part of pilot programs in Indonesia, participatory mapping was used to collect data for preparedness and flood contingency planning. This effort was led by the Province of Jakarta's Disaster Management Agency (BPBD) and involved 500 representatives from all 267 urban villages and 70 students from the University of Indonesia. OpenStreetMap tools were used to map 6,000 buildings and critical infrastructure, including schools, hospitals, and places of worship and 2,668 RW (subvillage) boundaries (see figure B3.1.1). Data can be analyzed in the InaSAFE tool for planning and preparedness (http://inasafe.org). As a follow-up, the Indonesian National Disaster Management Agency (BNPB) is extending the exercise to other hot spots around the country.

OpenStreetMap offers several important features for participatory mapping: Open source tools for online or offline mapping, a common platform for uploading, hosting data with free and open access and an active global community of users, and customized resources for a growing community in Indonesia; see http://en.openstreetmap.or.id/.

This work is part of an innovative approach to DRM through a partnership led by BNPB and AusAID through the Australia-Indonesia Facility for Disaster Reduction (AIFDR), UN OCHA Indonesia, and the Humanitarian OpenStreetMap team with support from the World Bank and Global Facility for Disaster Reduction and Recovery (GFDRR).

Figure B3.1.1 Illustration of OpenStreetMap, Jakarta

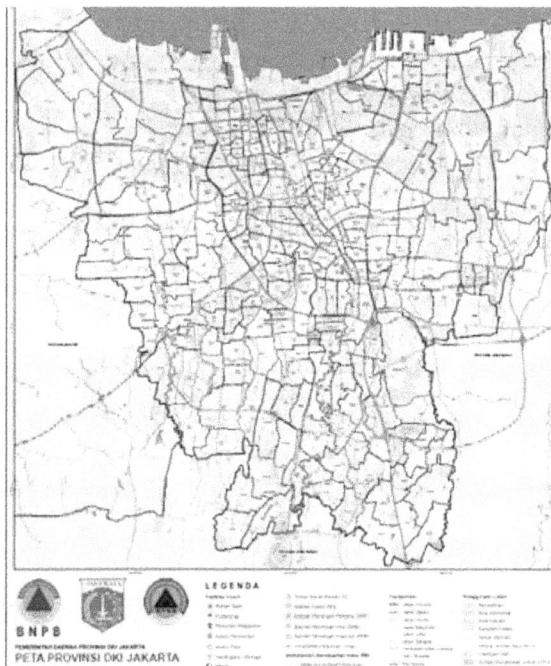

Source: World Bank staff.

Results of risk identification should be integrated across all core areas of DRM (see figure 3.5). To conduct robust risk identification, risk assessments of floods, tropical cyclones, and drought require a range of hydromet information. Investing in better national meteorological or hydromet services (NMSs and NHSs) will enhance a county's capacity to conduct risk identification (see also chapter 3). Availability of forecast hydromet information can be combined with scenario risk assessment tools to enable forecasted impact assessment of a given hazard. Knowing that hydromet hazards are the most expensive in East Asia and the Pacific, such information provides a powerful decision-making tool to anticipate scenarios of hazards and their forecasted impacts on assets and people. Integrating information from baseline and forecasted hydromet events will also greatly enhance early warning systems in a way that decision makers will be able to prepare people located in the affected area identified by the forecast, protect public assets when possible, and alert communities of imminent risk. Baseline data that feed into robust risk identification are also valuable for a risk-financing mechanism, providing accurate and quantitative information (see also chapter 6). If public assets are inventoried in advance as part of baseline data collection, governments will be able to limit their contingent liability to disasters by investing in retrofitting where needed or adapting new infrastructures to disaster-resilient standards.

Communicating risk and uncertainty can be challenging. Risk managers need to be realistic about limitations of predictions and the ability of risk mitigation and preparedness measures in fulfilling their role of reducing risk and damage or protecting from disaster impacts. These limitations as well as potential consequents of the failure of measures need to be communicated to the public in a way that individuals and communities can make informed decisions about their lives. At the same time, many factors contribute to shape people's decisions, so that when communicating disaster risk, strategies have to be developed to fit different stakeholder preferences and capacities (See CRED and USAID 2012 or Weber 2011).

Figure 3.5 Using Risk Assessment in Building Resilience

Risk assessment as a core part of DRM					
Risk assessment of communities likely to be impacted by cyclone	Earthquake hazard assessment	Understanding consequences based on a risk assessment of hurricane scenarios	Inundation extent of a flood for a certain period of time	Structural vulnerability assessment of buildings in the event of tsunami	National level probabilistic risk assessment

Actions taken					
Cyclone early warning system at community level	Retrofitting schools and hospitals	Land-use planning	Shelter construction and supplies for emergency response	Building selection and evacuation planning	Sovereign risk financing

What Needs to Be Done?

1. **Invest in deepening in-country technical capacity and scientific knowledge of disaster risks.** For sustainability and scalability, development partners and countries themselves should support the development of resources to conduct risk identification, especially at subnational levels. Large-scale disasters are more documented than small-scale localized disasters, although documentation of the former remains incomplete. However, small-scale, high-frequency events with accumulated impacts can lead to disastrous effects comparable to those from large-scale, rare, impressive events. Higher resolution data at a subnational scale are valuable information to inform local decision making, such as local land-use planning. Investment is needed in the in-country capacity to conduct and continually update the risk identification process. An example of subnational risk identification in Fiji is presented in box 3.2 and one of regional risk identification in box 3.3.

Box 3.2 Integrated Flood Management Pilot in Fiji

Exposed to tropical cyclones and earthquakes, flooding affects Fiji more often than any other natural hazard, with average annual losses per year of 10 lives and US$20 million in damages, and individual flood costs exceeding US$200 million. During the recent record flooding in 2009 and 2012, Nadi—the island's main entry point—was up to four meters under water. The 2009 and 2012 floods devastated Viti Levu's west coast, destroying bridges and roads, cutting off electricity, damaging houses, and disrupting lives and livelihoods as well as the sugar industry.

With the support from the World Bank and the Global Facility for Disaster Reduction and Recovery (GFDRR), the South Pacific Community–Applied Geoscience and Technology Division (SOPAC) is leading a pilot project to strengthen flood risk management in the Nadi basin by utilizing the best available disaster and climate information. In a timely and innovative approach using state-of-the-art technology and scientific models, including high-resolution LiDAR topography data and two-dimensional hydrodynamic modeling, the Nadi flood pilot assists local authorities and regional partners to gain a detailed understanding of the flood risks and possible mitigation options. As part of the project, high-quality hazard data have been collected, which feed into a detailed risk assessment and help to identify the most effective measures for flood risk mitigation and preparedness.

By providing evidence-based integrated flood management solutions, the initiative circumvents past experiences that focused on single measures that fell short of alleviating flood risks. Along with SOPAC, the World Bank/GFDRR, and local authorities, partners in the initiative include the Asian Development Bank, Australian Department of Climate Change, European Union, Global Environment Facility, and UN-Habitat. Based on an evaluation of the pilot's results, there are plans to replicate the approach in other flood-prone areas in Fiji.

Source: World Bank staff.

Box 3.3 Pacific Catastrophe Risk Assessment and Financing Initiative

The Pacific Catastrophe Risk Assessment and Financing Initiative (PCRAFI) is a joint program between the Secretariat of the Pacific Community (SOPAC), the World Bank, and the Asian Development Bank, with financial support from the government of Japan and the Global Facility for Disaster Reduction and Recovery (GFDRR) and the following participating countries: Cook Islands, Fiji, Kiribati, the Marshall Islands, the Federated States of Micronesia, Nauru, Niue, Palau, Papua New Guinea, Samoa, the Solomon Islands, Timor-Leste, Tonga, Tuvalu, and Vanuatu.

The initiative developed for the first time probabilistic and quantitative risk models and maps of tropical cyclone, earthquake, and tsunami risks covering the entire land mass of Pacific island countries (PICs). This platform provides a unique opportunity for PICs to integrate risk information into their development planning and decision-making processes. The geographic information system (GIS) database provides full coverage of the entire land mass of the select countries by means of field visits to survey more than 80,000 buildings, digitizing from satellite imagery the footprints of 450,000 buildings, and inferring from satellite imagery 2,900,000 buildings and other assets (see figure B3.3.1).

The Pacific Risk Information System developed as part of PCRAFI hosts probably the largest regional geospatial database and country-specific catastrophe risk models, with data and information on population, detailed asset information (buildings, infrastructure, and crops), and land cover, as well as historical catalogs and information on cyclones, earthquakes, and tsunamis. Exposure, hazard, and risk maps are accessible through this platform as powerful visual tools for informing decision makers, facilitating communication and education on DRM.

Figure B3.3.1 Illustration of PCRAFI: Field-Surveyed Bridge in Fiji with Photo Validation

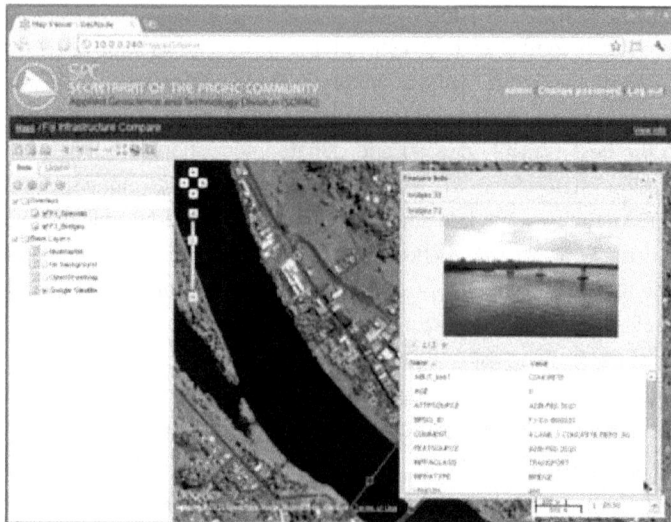

Source: World Bank staff.

box continues next page

Box 3.3 Pacific Catastrophe Risk Assessment and Financing Initiative *(continued)*

The platform enhances management and sharing of geospatial data within the Pacific community by creating a dynamic online community around risk data by piloting the integration of social web features with geospatial data management. Cost-effective development of the platform was achieved by using Geonode, an open source, web-based tool that allows sharing, management, and publication of geospatial data.

The risk information and tools developed under PCRAFI form the basis of applications on (1) developing disaster risk-financing and transfer strategies, (2) estimating quickly the extent and severity of disaster events, (3) predicting losses from tropical cyclones before they make landfall, (4) providing evidence for climate change negotiations in estimating impacts from projected future tropical cyclones, and (5) urban and infrastructure planning.

2. **Integrate risk information into decision making.** Key decisions, such as design of critical infrastructure or revision of land-use planning guidelines, should account for the existing and potential risks. Identifying critical infrastructure and vulnerable communities in a standardized format helps to target and prioritize investments. Baseline data are necessary for effective disaster response and as input into pre-disaster risk assessments, which can inform risk reduction investments. A systematic and participatory assessment of the risks and needs at the local and community levels, conducted not only post-disaster but also as part of pre-disaster risk identification, can greatly support the development of better-targeted programs. The social impact assessment, described in depth in chapter 7, could be used as a starting point for a more systematic collection of socioeconomic information on populations at risk before a disaster.

3. **Bridge the gap between national and community-level risk identification activities.** From government to industry, nongovernmental organizations to international organizations and communities at national, regional, and local levels can contribute to building collective resilience. National-level agencies are in the best position to fund and develop highly technical, scientific information on hazards and risk. Risk identification activities conducted at the local government and community levels will help in implementing risk reduction that is tailored to the specific needs of a given location and population. Participatory mapping at local and district levels provides untapped opportunities for communities' knowledge to be leveraged in the risk identification process. Promoting citizen participation in this process is a powerful tool in raising awareness of disaster risks thus building resilience. Communities often developed a coping capacity and adaptive behaviors to environmental hazards, including the exposure to disasters. Making good use of such information and empowering communities with tools to develop their own risk information by the means of local technology has a high-return potential for disaster risk reduction.

Strong, Safe, and Resilient • http://dx.doi.org/10.1596/978-0-8213-9805-0

4. **Create positive incentives for sharing information.** Sharing risk information with the relevant stakeholders by making it open and available in a range of data formats, through web-based tools and direct offline dissemination, enhances disaster preparation. First, it allows the use of data by the relevant stakeholder for evaluating existing or predictable risks in communities. Second, dissemination of hazard maps and publication of building codes based on scientific consideration increase public awareness and involvement in protecting themselves and their assets. Third, sharing risk information, collected through risk assessments, promotes transparency and accountability. Governments are in a strong position to set a positive example by sharing information in a transparent and accountable way, which can inspire similar commitments of openness from stakeholders in the private sector, academia, and civil society.

How Can the World Bank Help?

The World Bank provides technical assistance linking multihazard risk assessments with other core activities of DRM, including disaster risk financing. Geographically isolated, with often low technical and financial capacity, East Asia and the Pacific traditionally has been in a weak position to gather and sustain risk information. As part of the disaster risk-financing insurance efforts led by the World Bank, the Pacific Catastrophe Risk Assessment and Financing Initiative (box 3.3) provides a context for collecting risk information and baseline data at a regional level, resulting in the development of the largest collection of risk data available to all stakeholders.

The World Bank is committed to investing in innovative open source software tools for risk information needs that can be easily maintained and adapted to new contexts. Indonesia and countries in the Pacific started to explore the use of new technologies in DRM by developing the Pacific Risk Information System and InaSAFE (Indonesia Scenario Assessment for Emergencies). Such free and open-source GIS tools are now available to other countries and public users for efficient GIS data management and decision support (see also Harvard Humanitarian Initiative 2011). In 2011 the Indonesian Disaster Risk Management Agency, with the support of the Australia-Indonesia Facility for Disaster Reduction and the World Bank and Global Facility for Disaster Reduction and Recovery (GFDRR), developed a simple risk assessment tool, InaSAFE, that produces impact maps to guide key DRM decisions. The tool is operationalized by BNPB for province- and district-level contingency planning and is currently being customized for new use cases (see figure 3.6).

Scenarios help agencies better prepare for disasters by indicating the costs and benefits related to specific (mitigation) actions. An important feature of the InaSAFE tool is that it offers an impact analysis platform that combines scientific hazard scenarios such as the tsunami in Maumere, flooding in Jakarta, earthquake ground shaking in Padang, and a volcanic ash fall near Mount Merapi with localized building and population exposure data. The

Figure 3.6 Illustration of InaSAFE Output

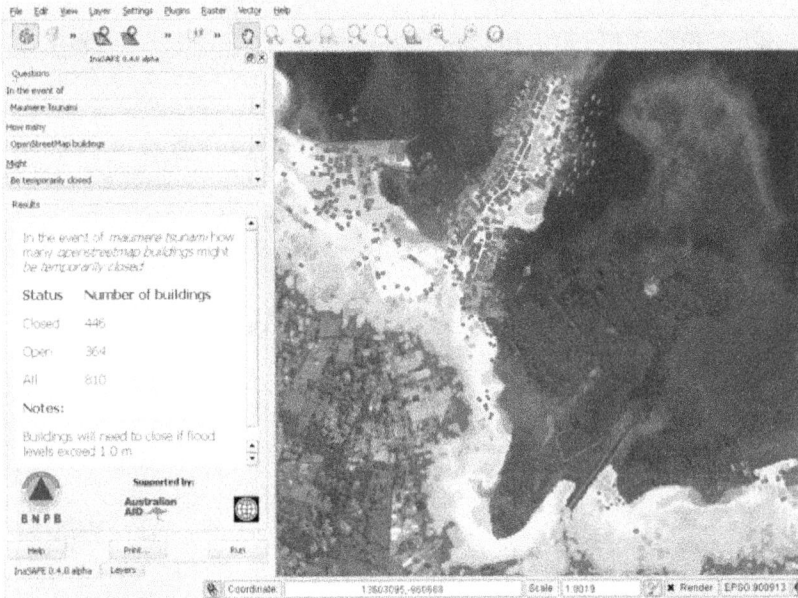

hazard scenarios are produced by technical experts who share data layers in a standardized format. For all the available use cases, specific scenarios, such as a rapid post-event response to earthquakes, pre-event preparedness, and early warning based on forecast hydromet information, can be quickly updated to reflect the latest impact results. Input data for exposure can easily be integrated from a range of sources, including participatory mapping exercises such as described in box 3.1. Based on the results of the impact analysis, the InaSAFE tool presents suggested actions to respond to the expected or forecasted impacts. Currently these actions are developed according to BNPB's operational guidelines and needs and would require customization for other country contexts.

Integrating satellite earth observation for high-resolution risk identification. The World Bank has an ongoing collaboration with the European Space Agency (ESA) to integrate satellite-derived information in DRM programs. The aim is to establish a scientific and practical link between DRM and new developments in earth observation (EO) initiatives, programs, and services. EO imagery allows the extraction of hazard information (flood risk area, subsidence, and landslide) and exposure (buildings, roads, dams) at very high resolution for a detailed local-level analysis. Through this program regional projects in Ho Chi Minh City, Jakarta, and Yogyakarta have already benefited from cutting-edge EO techniques (see figure 3.7). One of the most promising applications is interferometric synthetic aperture radar–based

Figure 3.7 Examples of EO Information Products

a. InSAR-based PSI in Jakarta

b. Flood hazard and risk data for Ho Chi Minh City

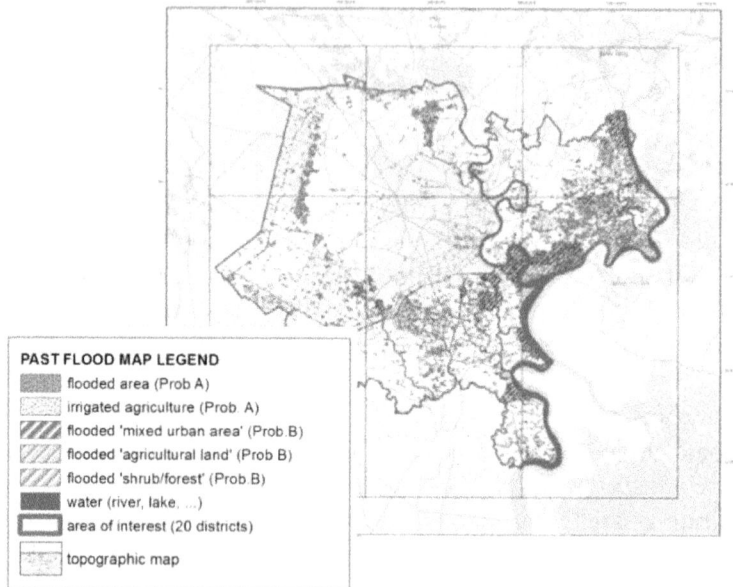

PAST FLOOD MAP LEGEND
- flooded area (Prob A)
- irrigated agriculture (Prob. A)
- flooded 'mixed urban area' (Prob.B)
- flooded 'agricultural land' (Prob B)
- flooded 'shrub/forest' (Prob.B)
- water (river, lake, ...)
- area of interest (20 districts)
- topographic map

Sources: Jakarta: Jakarta deformation map based on PSI analysis derived from VHR Cosmo-SkyMed data (October 2010–April 2011), from the Eoworld Project/Altamira Information for ESA and World Bank; Ho Chi Minh City: past flood map (October–November 2001) based on Radarsat-1 data, from the Eoworld Project/Eurosense for ESA and World Bank.

persistent scatterer interferometry (InSAR-based PSI), known for providing detailed measurements of surface displacements for measuring hazards associated with earthquakes, volcanoes, and landslides, as well as subsidence and deformations of flood defense structures in coastal lowlands. Another example is the use of EO for flood risk analysis. The integration of EO-derived information about past and potential flood events with estimation of land-use and asset exposure can support flood risk management decision making.

Sharing data reduces vulnerability to disasters by effectively communicating risk. The Open Data for Resilience Initiative (OpenDRI), led by GFDRR in partnership with the World Bank and other development institutions, aims to reduce the impact of disasters by empowering a wide range of stakeholders from policy makers to civil society with better information and the tools to support

---

Box 3.4 What Does It Mean for Data to Be Open?

In its simplest framing, data can be said to be open when both legally and technically open. For data to be considered legally open, data must be released under a license that allows for the reuse and redistribution for either commercial or noncommercial uses. Examples include the Creative Commons suite of licenses[a] or the Open Data Commons Open Database License.[b] Technically, open data are data available over the web on a permanent address, which can be downloaded or accessed through an application programming interface in structured and nonproprietary formats. Open formats for geospatial data, which comprise the majority of risk-related information, include shapefile, GeoTiff, and CSV or Open Geospatial Consortium (OGC) standard web services.

The open data philosophy has existed within the scientific community for decades, championed by those who argue that facts cannot be copyrighted and point to the many ways in which free access to basic data encourages beneficial research and innovation in academia and the private sector. More recently there has been a strong emphasis on open government data as part of a larger strategy means to promote transparency, accountability, and participation in governance. The Open Government Partnership[c] is a new multilateral initiative under which 47 governments have committed themselves to adopting these principles as part of anticorruption efforts, improving delivery of public services, and other endeavors. Web portals such as Data.gov have made enormous amounts of government data available for development institutions, and the World Bank, African Development Bank, and USAID have also adopted open data policies and practices in the last few years.

Additional resources:
- Open Data Handbook: http://opendatahandbook.org
- International Aid Transparency Initiative: http://www.aidtransparency.net
- OGC Standards: http://www.opengeospatial.org/standards.

Source: World Bank/GFDRR 2012.
a. http://creativecommons.org.
b. http://opendatacommons.org/licenses/odbl/.
c. http://www.opengovpartnership.org.

---

their decisions (see box 3.4). OpenDRI supports one of the key policy recommendations of the joint World Bank report, *Natural Hazards, UnNatural Disasters: The Economics of Effective Prevention* (World Bank 2010), and builds upon the World Bank's broader Open Data Initiative. The East Asia and Pacific Disaster Risk Management portfolio has multiple projects implementing the OpenDRI approach,[3] including PCRAFI (box 3.3), InaSAFE, and OpenStreetMap participatory mapping.

Notes

1. The most relevant indicators for risk identification include: 2.1 National and local risk assessments based on hazard data and vulnerability information are available and include risk assessments for key sectors; 2.2 Systems are in place to monitor, archive, and disseminate data on key hazards and vulnerabilities; 3.1 Relevant information on disasters is available and accessible at all levels, to all stakeholders; 3.2 Research methods and tools for multirisk assessments and cost benefit analysis are developed and strengthened. See also appendix D.

2. "The Compilation of National Progress Reports on the Implementation of Hyogo Framework for Action 2009–11," is valuable for assessing the current status in countries in East Asia and the Pacific.

3. http://www.gfdrr.org/gfdrr/opendri.

References

CRED and USAID. 2012. "CRED Crunch. Disaster Data: A Balanced Perspective." *CRED Crunch* 27 (February). http://reliefweb.int/sites/reliefweb.int/files/resources/PDF_116.pdf.

Harvard Humanitarian Initiative. 2011. *Disaster Relief 2.0: The Future of Information Sharing in Humanitarian Emergencies*. UN Foundation & Vodafone Foundation Technology Partnership. Washington, DC, and Berkshire, U.K.

UNISDR (United Nations International Strategy for Disaster Reduction). 2009. *Regional Synthesis Report on Implementation of the HFA in Asia and Pacific 2007–2008/09*. New York.

———. 2010. *Protecting Development Gains, Reducing Disaster Vulnerability and Building Resilience in Asia and the Pacific*. The Asia Pacific Disaster Report, New York.

———. 2011a. *HFA Progress in Asia Pacific: Regional Synthesis Report 2009-2011*. New York. http://www.unisdr.org/we/inform/publications/21158.

———. 2011b. *Hyogo Framework for Action 2005–2015: Building the Resilience of Nations and Communities to Disasters, Mid-Term Review 2011*. New York. http://www.preventionweb.net/files/18197_midterm.pdf.

———. 2011c. *The Compilation of National Progress Reports on the Implementation of Hyogo Framework for Action 2009–2011*. New York. http://www.preventionweb.net/english/hyogo/progress/documents/hfa-report-priority3-1(2009-2011).pdf.

Weber, E. 2011. "The State of Research on Behavior and Climate Applications." Paper presented at Garrison Institute, "Climate, Mind, and Behaviour Symposium," New York, March 2. http://research.iarc.uaf.edu/nx2020/slides/101021WeberRisk_1.pdf.

World Bank. 2010. *Natural Hazards, UnNatural Disasters: The Economics of Effective Prevention.* Washington DC: World Bank.

———. 2011. *Pacific Catastrophe Risk Assessment and Financing Initiative. Country Risk Profile: Papua New Guinea.* Washington, DC.

———. 2012. *Building Urban Resilience, Tools for Building Urban Resilience: Integrating Risk Information into Investment Decisions.* Pilot Cities Report, Washington, DC.

World Bank/GFDRR (Global Facility for Disaster Reduction and Recovery). 2012. "Understanding Risk: Best Practices in Disaster Risk Assessment." Proceedings from the 2012 UR Forum, Washington, DC.

CHAPTER 4

Risk Reduction: Measures and Investments

Key Messages for Policy Makers

- Invest in prevention: Rising exposure does not have to automatically translate into increasing risks if preventive measures and approaches are embedded in the design and construction development investments exposed to disaster risks.
- Risk reduction is not a department that can stand alone: Its multidisciplinary nature requires disaster risk management (DRM) efforts to be mainstreamed into sectors at risk, and coordinated by a high-level ministry to enforce and monitor implementation.
- The balance between engineered and nonstructural solutions is crucial: Institutional arrangements that facilitate DRM integration across sectors, land-use regulations, enabling policies, better ecosystem management, risk awareness, and a stronger focus on social protection and community-driven development programs are equally important as investments in risk-reducing infrastructure.
- Nonstructural measures can be highly cost-effective: These also often represent a no-regret flexible approach to reducing and managing disaster risk.

Where Are We Now?

In rapidly urbanizing East Asia and the Pacific, disaster risks are increasingly an outcome of development processes. Unfortunately, too often urban land-use plans and risk reduction strategies are developed in isolation of one another. Development programs do not necessarily reduce vulnerability to natural hazards. Instead, they often unwittingly create new forms of vulnerability or exacerbate existing ones, sometimes with tragic consequences, for example, through building in hazard-prone zones or failing to apply disaster-resilient building

---

This chapter was written by Henrike Brecht with input from Zuzana Stanton-Geddes.

codes. Much of the current development practice and programming in East Asia and the Pacific fails to address risk reduction. As a result, natural structures, for instance, wetlands, mangroves, dunes, and flood plains, that form natural buffers between people and nature are being eliminated in the quest for growth, leaving people and assets highly exposed to disasters (box 4.1).

Most national disaster systems in East Asia and the Pacific are still reactive in their actions, with the majority of disaster spending allocated after instead of before a disaster occurs. Often linked to skewed incentives and overreliance on post-disaster aid, this attitude poses a serious challenge for mainstreaming disaster risk reduction into urban, social, economic, environmental planning, and development. In most countries in East Asia and the Pacific, risk reduction is not addressed throughout sectoral frameworks and institutional structures, country strategies, and policies and in the design of projects. A range of factors hinder effective mainstreaming, including weak engagement by the development sectors, limited authority of current national disaster management organizations to require sectors to include risk reduction measures, weak legal frameworks and policies, lack of funding, and difficulties in implementation and accountability, as summarized in chapter 2.

Countries traditionally rely on engineered solutions that can become obsolete in the context of rapid urban growth and climate change uncertainties. Despite progress in taking a balanced approach to disaster prevention, governments in East Asia and the Pacific still have the tendency to concentrate on hardware solutions, neglecting policies, planning, and institutions to achieve cost-effective, efficient, and participatory risk reduction. In the case of flooding, for instance, physical flood defenses can address only an element of the issue at stake. Instead of first assuming that more embankments and pipes are the answer, a more holistic environmental approach could be considered, including options such as wetlands restoration. Just as important is the willingness to preserve natural capital and relocate or limit urban and industrial expansion away from high-risk land, such as low-lying coastal zones. As a city

---

Box 4.1 Building in Harm's Way

An example for rapid development in vulnerable locations is the Ayutthaya Province in Thailand, where industrial parks expanded on former swamps that used to yield good quality rice precisely because of regular flooding. The 2011 floods overwhelmed the six-meter-high levees built to protect these estates. As a result 891 factories in industrial estates that employed about 460,000 people were closed. The country-wide fatality count in the country exceeded 800, and a World Bank study estimated the total losses at US$46 billion. Insured losses were estimated to reach US$12 billion—the highest number on record for a single flood event, according to Swiss Re.

Sources: Thai Industrial Estate and Strategic Partners Association and World Bank 2012d.

---

develops, large-scale engineered solutions such as flood protection schemes can face challenges even before they are completed (see table 4.1 for specific sectors). This was the case in Ho Chi Minh City, Vietnam's rapidly growing commercial center built in a low-lying flood area. The 2001 Master Plan intended to mitigate flooding through improved drainage but had to contend with higher-than-expected increases in peak rainfall before individual measures were implemented.

Table 4.1 Sectors Where Inertia (Lock-Ins) and Sensitivity to Climate Change Are Great

Sector	Example	Time scale (years)
Water	Dams, reservoirs	30–200
Land-use planning	Mew development in flood plains or coastal areas	>100
Coastal and flood defenses	Dikes, sea walls	>50
Building and housing	Insulation, windows	30–150
Transportation	Port infrastructure, bridges, roads, railways	30–200
Urbanism	Urban density, parks	>100
Energy production	Coal-fired plants	20–70

Source: Hallegatte 2009 in World Bank 2012c.

Where Do We Want to Be?

"Anticipate and prevent" instead of "wait and see." As described in chapter 1, governments in East Asia and the Pacific have made advances in the areas of strengthening capacities, institutional systems, and legislation, particularly to address shortcomings in the areas of disaster preparedness and response. Effective disaster prevention relies on a shift away from traditional disaster response toward multisectoral risk reduction cooperation with stakeholders across different government levels as well as the private sector and communities at risk. Disasters should not be considered as inevitable, temporary disruptions, which can be managed reactively and irregularly through humanitarian response and reconstruction, but rather as events that require a more proactive approach that reduces the costs of hazards before disaster events. This shift requires a strong investment in national capacities for governments to lead and implement comprehensive risk reduction agendas and to coordinate between ministries and different stakeholders.

Disaster-prone areas can reap large benefits from risk reduction measures. Disasters in 21 countries in Africa, Asia, and Latin America have damaged or destroyed 63,667 schools since 1989. Forty-six percent of these schools were damaged or destroyed in frequently occurring disasters rather than in occasional and large catastrophes (UN 2011). This large amount of damage leads to an unacceptable loss of children's and teachers' lives, extensive losses in government budgets, and a reduction of enrollment rates. Investing in early warning systems, for example, is a preparedness measure that pays off. In some countries, the enhancement of early warning systems has led to striking results in reducing mortality risk, such as in Hong Kong SAR, China (UN 2011). Building resilient critical infrastructure, in particular safe schools, is a risk reduction priority. Because it is not cost-effective to retrofit all schools at risk, prioritization methods are applied that show the highest cost-benefit ratios. Construction standards and building codes need to match the level of risk. See box 4.2 for further information about country-wide earthquake management program and appendix E for an action plan for building earthquake resilience.

Making decisions on the choice of measures, risk assessments, and cost-benefit analyses helps to define a pragmatic mix of instruments depending on a country's capacity and available funds (see also chapter 2). The cost-benefit analysis calculates where maximum gains in risk reduction can be made and includes the identification of a scenario with and without risk reduction intervention, quantification of the impacts in both scenarios, and calculation of the costs and benefits over the lifetime of a given investment. Measures that bring benefits under a range of scenarios are important when dealing with disaster risks and uncertainties related to rapid urbanization, growth patterns, or climate change impacts. A robust decision-making process can help countries determine low-regret strategies that are cost-effective in the long run. Box 4.3 illustrates how Vietnam is dealing with risk and uncertainties in Ho Chi Minh City. This approach underlines that disaster risk management is an iterative process; the

---

Box 4.2 What Countries in East Asia and the Pacific Can Do to Prepare for the Next Big Earthquake

History, observations of damage after strong earthquakes, and engineering assessments and analyses all find that the following public buildings and infrastructure and their key nonstructural features and equipment are highly vulnerable and should or could be addressed first:

- **Schools, hospitals, and critical government buildings,** such as firehouses and police stations and other buildings needed for emergency response.
- **Public infrastructure,** including key highways and bridges, airports, electric power generation and distribution systems, water and wastewater systems, and telecommunications systems.

Country-wide earthquake risk management programs involve risk assessments, followed by multiphased risk reduction programs that can take from a few years to decades to complete. Such programs have been successfully carried out in several countries. The programs typically consist of three phases:

1. **Risk audit of a specific sector,** such as public schools. This should be a quick study based on experience and very limited engineering analyses.
2. **Detailed risk assessment,** including cost-benefit analysis for the particular sector.
3. **Implementation,** that is, reducing the risk through strengthening and renovation of the structures and bracing their important equipment and nonstructural components.

Source: World Bank 2010. See also Appendix E.

Box 4.3 Dealing with Uncertainties: Experience from Ho Chi Minh City, Vietnam

As a city develops, large-scale flood protection schemes often face new challenges even before they are completed, as, for example, in Ho Chi Minh City, where the 2001 Master Plan to mitigate flooding through improved drainage had to contend with higher than expected increases in peak rainfall. Currently the Ho Chi Minh City Steering Center of Urban Flood Control is preparing an Integrated the Flood Management Strategy to synchronize the existing master plans regarding the storm sewer system, flood control system, and space development through 2025 and to adapt Ho Chi Minh City to climate change. These efforts are also a response to increases in precipitation and tide levels observed over the last decade already exceeding those projected. Unanticipated changes raise concerns that the original plans may not manage flooding in the city and could even make it worse in some areas. The strategy will be decided through a robust decision support system framework.

Source: World Bank 2012b.

---

decisions taken today should allow countries to adapt, should conditions and needs change.

A blend of hard and soft measures is critical to reduce risk because structural measures can prove unsustainable under large hydrological, land subsidence, and urbanization uncertainties. Hard or gray measures include, for example, investments in infrastructure, from levees and dams to retrofitting of critical buildings. Flood control reservoirs, for instance, collect water in times of heavy rainfall and then release it slowly over the course of a longer time period. Soft measures include institutional arrangements, land-use regulations, public education, social protection and community-driven development, and DRM interventions, as well as the provision of economic incentives to promote a risk-based approach to development. Delineating flood zones in land-use plans and issuing policies to restrict development in these zones is an example of a soft measure. More countries are taking a balanced approach, with the Netherlands being one of the forerunners. The Dutch Room for the River program moves dikes inland and deepens riverbeds on a large scale to create more space for aquatic systems. In East Asia and the Pacific, Vietnam is taking a more balanced approach by restoring its coastline, whereas in Jakarta and Manila, flood mitigation plans are looking at integrating nonstructural measures into their strategies. Similarly, following the great east Japan earthquake and tsunami, the government of Japan is actively promoting a balanced strategy.

Shifting from engineered infrastructure (gray) solutions to a balance of gray and green defense mechanisms has shown to be effective in terms of outcomes and saving costs. Green infrastructure measures such as rain gardens, bioswales, permeable pavements, and urban green spaces provide co-benefits in the form of improved streetscapes, provision of local jobs, reduction of the heat island effect, and improved air quality, among others. However, codes and regulations still need to be modified to allow the use of green infrastructure in lieu of traditional gray methods. Over time, as the approaches are used on a larger scale and in different areas of the world, more information should be gathered to determine whether these measures are robust and perform adequately in the long term. Guidelines can then be developed to inform the use of green infrastructure methods and enable their application to be scaled up where appropriate. Box 4.4 offers examples of cities that adopted a green infrastructure approach.

Box 4.4 Cities' Experience with a Green Infrastructure

In New York City, modeling showed that a green strategy will reduce more storm water volumes at significantly less cost to New Yorkers than the all-gray strategy previously contemplated (figure B4.4.1). The green infrastructure option builds on the cost-effective gray infrastructure but also includes investments such as stream buffer restoration, green roofs, and bioswales, whereas the gray solution concentrates solely on human-engineered tanks,

Figure B4.4.1 New York City-wide Costs of Combined Sewer Overflow Control Scenarios after 20 Years

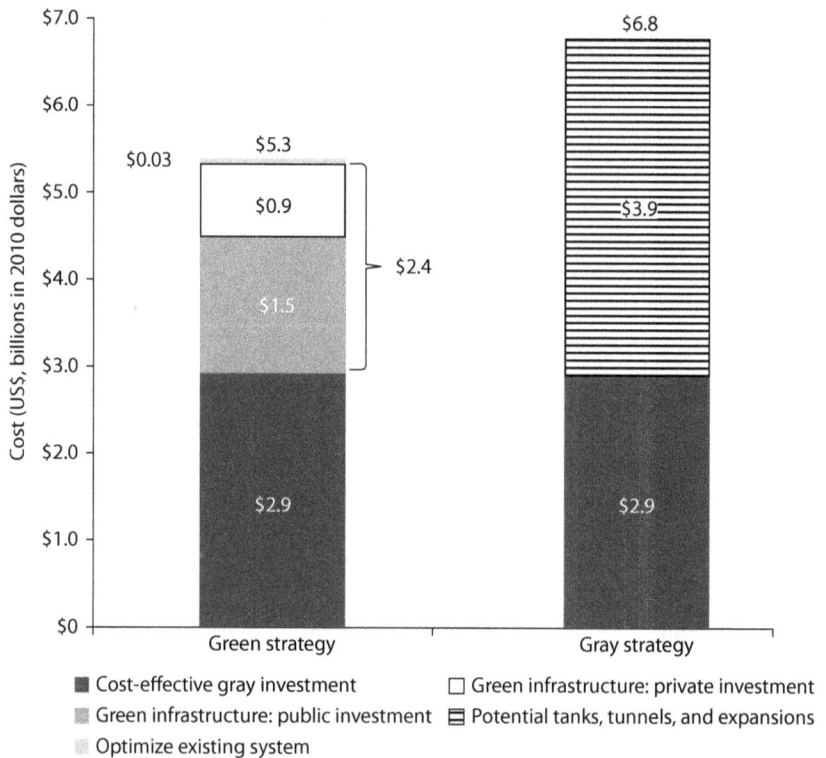

Cost (US$, billions in 2010 dollars)

Green strategy: $0.03, $5.3, $0.9, $1.5, $2.9, $2.4

Gray strategy: $6.8, $3.9, $2.9

■ Cost-effective gray investment
▨ Green infrastructure: public investment
▧ Optimize existing system
□ Green infrastructure: private investment
☰ Potential tanks, tunnels, and expansions

Source: Adapted from City of New York 2010.

box continues next page

Box 4.4 Cities' Experience with a Green Infrastructure *(continued)*

tunnels, and storm drains. Cost savings with the green infrastructure plan total more than US$1.5 billion (City of New York 2010).

In Seattle, residents are reimbursed for installing "rain gardens," which are designed using native plants and special soil to reduce rainwater runoff, instead of allowing infiltration into the ground. Storage of rainwater for future use in watering lawns and gardens is also encouraged.

In Chicago, permeable pavement and "cool" pavement are being used in alleys to increase urban rainwater infiltration and decrease the heat island effect from conventional paving material.

In Enkoping, Sweden, phyto-remediation has been used to treat sewage by pumping sludge onto 190 acres of coppiced willow trees. These trees filter out the pollutants in the sewage, and, when harvested, the willows are used as biofuel to generate electricity.

Sources: Authors and City of New York 2010.

Using existing social protection and community-driven development interventions can be particularly effective, especially for smaller scale disaster preparedness investments. Including DRM elements into social protection and community-driven development programs, described in detail in chapter 2, has the potential of substantially reducing disaster response costs by drawing on a preexisting network of case workers and community facilitators and on already functioning systems to deliver support to households. These types of approaches are consistently cost-effective (because they are able to save on contractors' profits). In the Philippines cost savings ranged from 8 percent for school buildings to 76 percent for water supply investments when compared with traditionally implemented infrastructure (Araral and Holmemo 2007).

What Needs to Be Done?

Although no single disaster reduction model works for all and strategies will vary across countries, measures exist that have been proven to be solutions with high cost-benefit ratios. External assistance can provide financing in the form of lending and grants. It can also help to derive innovative strategies, for example, through panels of highly respected experts who can advise on the best ways forward. Instruments that are useful for decision makers include risk maps, economic cost-benefit analysis, impact evaluations, and climate forecasts. As mentioned in chapter 2, partnerships have proven critical for successful risk reduction initiatives. Strong coordination and collaboration between different levels of government are especially important because they allow for local strengths while acknowledging that local government have limited resources.

1. **Get the balance in financing right.** The portion of the disaster budget spent on relief and repair often by far outweighs the fraction spent on prevention.

Strong, Safe, and Resilient • http://dx.doi.org/10.1596/978-0-8213-9805-0

This holds true for both government and donor spending. The U.S. federal government, for example, spent US$3.05 billion on disaster response versus just US$195 million on disaster prevention annually from 1985 to 2004 (Healy and Malhotra 2009). Yet the level of government preparedness greatly determines the extent of suffering and loss. Costs for prevention can be reduced through addressing risk factors early on, developing participatory community approaches, and combining it with regular infrastructure development. For example, school buildings and other public infrastructure in hazardous areas can be built to cyclone norms so that they can be used as shelters.

2. **Minimize the consequences of poor or unplanned urbanization by bridging risk reduction and urban planning through the right balance of structural and nonstructural measures.** Governments can minimize losses by factoring risk reduction into development. Integrating disaster risk reduction is especially important in key economic sectors and in sectors that have the highest losses due to disasters. Getting the right balance between infrastructure investments and nonstructural measures may include a stronger focus on improving institutional arrangements, regulations and coordination, sharing risk information (see chapter 3), and investing in systems, including social protection and community-driven development programs, that can allow a rapid outreach to households and communities (see chapter 2). Strengthening hazard forecast and hydromet services is a no-regret investment with a high cost-benefit ratio (see World Bank 2012a, and chapter 5). Restoring natural ecosystems can also be more cost-effective than engineered solutions. The World Bank flagship report on flood risk management can help countries in East Asia and the Pacific in selecting and implementing the right choice of measures when dealing with the challenge of urban flooding (box 4.5).

3. **Enforce multisectoral responsibilities and strong central coordination.** The department in charge of response and relief many times is ill-equipped to provide guidance on mitigation measures and investments. To mainstream disaster risk reduction into sectors and line ministries, three fundamental steps are needed: First, sectoral risk assessments need to be developed. For example, in the transport sector, this would translate to mapping vulnerable road stretches. Second, technical guidelines to address the identified vulnerabilities must be specified. This includes, for example, relocation of roads to higher ground, larger culverts, and bioengineering solutions for slope stabilization. And, third, awareness raising and training is needed for all levels of government but also, for instance, for road engineers and construction workers. To enforce and monitor the implementation of risk reduction initiatives, a high-level ministry will need to provide the coordinating guidance, overarching policies, and monitoring mechanisms across government. Some countries may require two different types of agencies—a high-level coordinating agency for

Box 4.5 Guiding Principles for Integrated Urban Flood Risk Management

1. Every flood risk scenario is different; there is no flood management blueprint.
2. Designs for flood management must be able to cope with a changing and uncertain future.
3. Rapid urbanization requires the integration of flood risk management into regular urban planning and governance.
4. An integrated strategy requires the use of both structural and nonstructural measures and good metrics for "getting the balance right."
5. Heavily engineered structural measures can transfer risk upstream and downstream.
6. It is impossible to entirely eliminate the risk from flooding.
7. Many flood management measures have multiple co-benefits over and above their flood management role.
8. It is important to consider the wider social and ecological consequences of flood management spending.
9. Clarity of responsibility for constructing and running flood risk programs is critical.
10. Implementing flood risk management measures requires multistakeholder cooperation.
11. Continuous communication to raise awareness and reinforce preparedness is necessary.
12. Plan to recover quickly after flooding and use the recovery to build capacity.

Source: World Bank 2012b.

policy mainstreaming, and a dedicated disaster response agency that can fall within an appropriate ministry.

4. **Consider disaster and climate change risks within a robust decision-making process.** Although rapid growth of assets and people in hazardous areas is the single biggest driver of risk (IPCC 2012), impacts of climate change can materialize in the future through increasing variability and extreme events. Good climate change adaptation (CCA) starts with effective DRM. As a first step, integration of DRM and CCA institutions is needed in countries where institutional duplication threatens effectiveness of action. In terms of processes, a robust approach to decision making, considering changing environments and climate uncertainties, can help in identifying a low-regret DRM strategy.

How Can the World Bank Help?

Including risk reduction into development that can help minimize disaster losses. In East Asia and the Pacific, the World Bank is supporting a range of preventive activities, such as helping Indonesia and the Philippines to identify the most critical schools and retrofit them. In Can Tho and Jakarta, the Building Urban Resilience Program, funded by the AusAID East Asia

Infrastructure Growth Fund, is helping to increase the resilience of urban infrastructure through city-level investments. Many current infrastructure investments automatically build in disaster risk reduction. For example, the Western Indonesia National Roads Improvement Project and the Vietnam Second Northern Mountains Poverty Reduction Project have components that build in resilience against disasters. Strengthening forecast and early warning systems is a no-regret measure with a traditionally high cost-benefit ratio (see also chapter 5).

Integrating disaster risk reduction, which is especially important in key economic sectors at risk and in sectors that have the highest losses due to disasters. The World Bank is supporting governments to mainstream risk reduction into investments. For example in the Lao People's Democratic Republic, for key sectors, including agriculture, transport, and urban planning, sectoral risk assessments are being conducted, identifying critical infrastructures at high risk. Based on these assessments, new guidelines and specifications are developed to make public investments resilient from disasters. Although this often requires slightly higher upfront investment, cost-benefit analyses of life-cycle costs can determine the return rate for these investments. Government staff and engineers are being trained in the new guidelines, and the mechanism is institutionalized by incorporating it into ministerial business processes. In areas at risk, it is more cost-effective to strengthen existing school buildings than to entirely rebuild them. The World Bank is also supporting Indonesia and the Philippines to identify the most critical schools and retrofit them.

Helping to implement the right balance between structural and nonstructural measures. Restoring nature is cost-effective because of the multiple benefits and long-lasting effects (See, for example, Dedeurwaerdere 1998; Kay and Wilderspin 2002; Tidwell 2005; Wells, Ravilious, and Corcoran 2006). Mangrove forests, for example, support fisheries by providing breeding grounds, they lessen the impact of toxic substances in water and soil, and they serve as a buffer against floods and typhoons. The World Bank's Vietnam Coastal Wetlands Protection Project planted 370 million trees along 460 kilometers of coast. By project close, erosion had been reduced by as much as 40 percent, and the area of coastline accretion had increased by 20 percent. Box 4.5 shows a selection of opportunities on how to reduce risk.

Promoting risk-sensitive land-use planning and resettlement. After the 2006 earthquake and the 2010 volcanic eruption in Yogyakarta, Indonesia, the governments and communities considered the reconstruction as an opportunity to rebuild safer settlements. Resettlement was offered to populations at risk as a last-resort measure—when risks could not be sufficiently mitigated through other means. The communities underwent a consensus-building process involving all stakeholders (community, nongovernmental organization, government, private sector, traditional leaders). They were educated about the risks and given the option of different resettlement schemes. Resettlement was not mandatory but remained voluntary for the communities.

Sharing risk information among stakeholders to strengthen collective resilience. The World Bank and Global Facility for Disaster Reduction and Recovery (GFDRR) have been supporting at-risk countries to reduce their risk through cutting-edge lending, technical assistance, and knowledge products. Examples of the World Bank's strategic work are described in chapter 3 and include developing open-source risk assessment platforms, making sectoral investments tools for risk reduction, and using land-use planning to reduce risks. Risk assessments are important disaster and climate-risk management tools for identifying risk, quantifying the potential impacts, and prioritizing mitigation measures. An example is the Pacific Catastrophe Risk Assessment and Financing Initiative (PCRAFI), illustrated in box 3.3. Open to all users, the PCRAFI data can inform government and donor projects related to macroeconomic planning, disaster risk financing, urban investments, infrastructure planning, and rapid post-disaster damage estimation. These approaches and tools are adaptable to other regions of the world.

Supporting DRM and CCA synergies through investments. The Bank's approach to small capacity-constrained states, such as Kiribati, Papua New Guinea, the Solomon Islands, and Vanuatu, carefully balances community resilient investment programs with building DRM and CCA institutions at national and subnational levels. Learning from past experience when highly fragmented investments could not demonstrate significant results, this approach is based on the principles of (1) integrating DRM and CCA under the banner of resilient development, (2) pooling of funding to avoid overstretching already stretched institutions, (3) demonstrating action on the ground, and (4) building the absorptive capacity of Pacific island countries to accommodate increased future climate financing.

References

Araral, E., and C. Holmemo. 2007. "Measuring the Costs and Benefits of CDD: The KALAHI-CIDSS Project Philippines." World Bank Social Development Paper 102, World Bank, Washington, DC. http://siteresources.worldbank.org/EXTSOCIALDEVELOPMENT/Resources/244362-1164107274725/3182370-1164201144397/3187094-1173195121091/SDP-102-Jan-2007.pdf.

City of New York. 2010. *NYC Green Infrastructure Plan: A Sustainable Strategy for Clean Waterways.* New York.

Dedeurwaerdere, A. 1998. "Cost-Benefit Analysis for Natural Disaster Management: A Case-Study in the Philippines." Centre for Research on the Epidemiology of Disaster Working Paper 143, Université Catholique de Louvain, Louvain-La-Neuve, Belgium.

Hallegatte, S. 2009. "Strategies to Adapt to an Uncertain Climate Change." *Global Environmental Change* 19 (2): 240–47.

Healy, A., and N. Malhotra. 2009. "Citizen Competence and Government Accountability: Voter Responses to Natural Disaster Relief and Preparedness Spending." http://myweb.lmu.edu/ahealy/papers/healy_prevention_070808.pdf.

IPCC. 2012. *Managing the Risks of Extreme Events and Disasters to Advance Climate Change Adaptation.* Special Report of Working Groups I and II of the Intergovernmental Panel on Climate Change. Edited by C. B. Field, V. Barros, T. F. Stocker, D. Qin,

D. J. Dokken, K. L. Ebi, M. D. Mastrandrea, K. J. Mach, G.-K. Plattner, S. K. Allen, M. Tignor, and P. M. Midgley. Cambridge, U.K.: Cambridge University Press.

Kay, R., and I. Wilderspin. 2002. "Box 4.4: Mangrove Planting Saves Lives and Money in Vietnam." In *World Disaster Report Focus on Reducing Risk*, 95. Geneva: International Federation of Red Cross and Red Crescent Societies (IFRCRCS). http://www.ifrc.org/Global/Publications/disasters/WDR/32600-WDR2002.pdf.

Tidwell, M. 2005. "Goodbye, New Orleans: It's Time We Stopped Pretending." AlerNet online Article 5, December 2005. Accessed September 2012.

UN (United Nations). 2011. *United Nations Global Assessment Report on Disaster Risk Reduction*. New York: United Nations.

Wells, S., C. Ravilious, and E. Corcoran. 2006. *In the Front Line: Shoreline Protection and Other Ecosystem Services from Mangroves and Coral Reefs*. Cambridge, U.K.: UNEPA World Conservation Monitoring Centre.

World Bank. 2010. *It Is Not Too Late: Preparing for Asia's Next Big Earthquake, with Emphasis on the Philippines, Indonesia, and China* [Policy Note], by P. I. Yanev. Washington, DC: World Bank.

———. 2012a. "A Cost-Effective Solution to Reduce Disaster Losses in Developing Countries: Hydro-meteorological Services, Early Warning, and Evacuation." Policy Research Working Paper 6058, World Bank, Washington, DC.

———. 2012b. *Cities and Flooding: A Guide to Integrated Urban Flood Risk Management for the 21st Century*, by A. Jha, R. Bloch, and J. Lamond. Washington, DC: World Bank. http://www.gfdrr.org/gfdrr/urbanfloods.

———. 2012c. *Inclusive Green Growth: The Pathway to Sustainable Development*. Washington, DC: World Bank.

———. 2012d. *Thai Flood 2011: Rapid Assessment Report for Resilient Recovery and Reconstruction*. Thailand: World Bank.

CHAPTER 5

Emergency Preparedness: Weather, Climate, and Hydromet Services

Key Messages for Policy Makers

- Weather, climate, and hydrological information is vital to minimize the growing economic losses from natural hazards, facilitate the adaptation to climate change, and guide economic development in sectors including agriculture, water resources management, transport, and energy production.
- Most weather, climate, and hydrological information is delivered by National Meteorological (NMS) and National Hydrometeorological (NHS) Services, which in many countries in East Asia and the Pacific are not able to address the current and future needs and mitigate natural hazards.
- Strengthening NMSs is needed to improve early warnings and weather and climate service delivery.
- Modernization of NMSs in developing countries is a high-value investment, providing a positive return to the national economy while improving public safety and security. In East Asia and the Pacific, China has shown that the cost-benefit ratio of strengthening NMSs can range between 1:35 and 1:40.
- Large efficiency gains in delivering weather, climate, and hydrological information can be made by developing regional meteorological or river basin hydrological systems that are better integrated within the global system.

Where Are We Now?

National Meteorological Services (NMSs) and National Hydrometeorological Services (NHSs) are the national institutions responsible for observations, forecasts, and warnings of extreme meteorological and hydrological events. In many countries in East Asia and the Pacific, the capacity of NMSs and NHSs to provide effective forecasts and warnings is inadequate. The main reasons are low visibility and inadequate attention from the government to these public sector agencies. Many governments do not understand the value of information and services,

This chapter was written by David Rogers and Vladimir Tsirkunov.

Strong, Safe, and Resilient • http://dx.doi.org/10.1596/978-0-8213-9805-0

which NMSs should provide as their public service mission. The existing poor status of an NMS prevents the production of valuable data and information. Governments often see no reason for investment in NMSs; without investment and support, there are no new products or services. In most cases the international support to NMSs has not been successful because of the limited scope of investments and capacity building, lack of coordination among donors, inadequate attention to sustainability of investments, limited duration of support, and other shortcomings.

Investments in strengthening NMSs and NHSs are cost-effective. Funding for national investment in hydrological and meteorological services is the first step to ensure that there is the capacity to deliver timely weather-, water-, and climate-related services for civil protection and economic development. In East Asia and the Pacific, China, for example, has shown that the cost benefit of NMS strengthening can range between 1:35 and 1:40 (Zhang and Haixiao 2003).[1] According to M. Jarraud, World Meteorological Organization (WMO) Secretary-General, "traditionally, the overall benefits accrued from investment made in the meteorological and hydrological infrastructures were estimated to be, in several countries, in order of 10 to 1" (Jarraud 2007). Studies undertaken in countries such as Switzerland and the United States have illustrated the high economic returns of NMS improvement, with cost-benefit ratios of 1:4 to 1:6 (Lazo, Morss, and Demuth 2009). Hallegatte estimates the potential benefits of upgrading all developing country hydrometeorological (hydromet) information production and early warning capacities to developed-country standards (Hallegatte 2012). Total benefits were estimated to be between US$4 billion and US$36 billion per year globally, with cost-benefit ratios between 4 and 36.

East Asia and the Pacific is dominated by monsoons and tropical cyclones,[2] which result in heavy rainfall, high winds, and periods of extended drought when the monsoons fail. While providing most of the precipitation that supports rain-fed agriculture, failure of the monsoons in onset and intensity can have devastating consequences in terms of food security. The majority of the more than 500 million rural poor in the region are subsistence farmers depending on rain-fed agriculture dominated by the monsoons, highly sensitive to climate variations. The monsoons largely determine the annual distribution of rainfall. In Vietnam, for example, the annual rainfall is between 1,800 and 2,500 millimeters with about 70 percent occurring in the summer monsoon season between May and September/October. In the Philippines, the average annual rainfall ranges from 5,000 to less than 1,000 millimeters in some sheltered valleys with the maximum also occurring during the summer monsoon. The Pacific island countries (PICs) are especially vulnerable to tropical cyclones and the resulting storm surges and flash floods because of their relatively low adaptive capacity. PICs rely on subsistence agriculture and fishing as means of livelihood, vulnerable to external shocks, including severe weather events. PICs are also at risk from extended periods of drought.

East Asia and the Pacific accounts for about 40 percent of the total number of floods worldwide over the past 30 years. As described in chapter 1, with the transition of low- and middle-income countries in the region into largely urban societies, loss of life and property due to the cyclone-related flooding, storm surges, and high winds are increasing. Floods and seasonal river flooding occur throughout the region with risk of storm surges along many coastlines. Tropical cyclones are the most costly meteorological disasters affecting East Asia and the Pacific with, on average, 27 tropical cyclones affecting some part of the region each year (Chan 2008)[3] (box 5.1).

Box 5.1 Rainfall in Asia in 2011

The entire region is the most active area for tropical cyclones, with 20 or more making landfall somewhere there. Heavy rainfall associated with the Asian summer monsoon resulted in above-average precipitation over the entire Indochina Peninsula in 2011. Flooding occurred over a wide area in the basins of the Chao Phraya River and the Mekong River. Serious damage occurred, especially in Thailand.

The four-month total precipitation from June to September 2011 was 120–180 percent of normal for most meteorological observation stations on the Indochina Peninsula. Four-month total precipitation for the period amounts to 921 millimeters (134 percent of the normal) at Chiang Mai in northern Thailand, 1,251 millimeters (140 percent) at Bangkok, 1,641 millimeters (144 percent) at Vientiane (the Lao People's Democratic Republic), and 835 millimeters (107 percent) at Phnom Penh (Cambodia). The most likely cause of the heavy precipitation was above normal cumulus convection during the monsoon.

Capacity of NMSs and NHSs in East Asia and the Pacific

In many countries, the capacity of NMSs and NHSs to provide effective forecasts and warnings is inadequate. NMSs and NHSs are the national institutions responsible for observations, forecasts, and warnings of extreme meteorological and hydrological events. Despite the importance of flood forecasting, cooperation between NMSs and NHSs is often weak, limiting advances in flash flood forecasting guidance and understanding and predicting the impact of severe floods.

The capacity to provide these essential services varies widely in East Asia and the Pacific.[4] The region includes some of the most advanced NMSs (Australia; China; Hong Kong SAR, China; Japan; the Republic of Korea; and New Zealand) as well as some of the least developed ones, which do not have the capacity to provide stand-alone services. Advanced NMSs in France,[5] India, the Russian Federation, the United Kingdom, and the United States provide support to NMSs in East Asia and the Pacific bilaterally or through the WMO. In the region, the China Meteorological Administration (CMA) is particularly advanced in the provision of Multi-Hazard Early Warning Systems (MHEWSs), with an example of a best practice operated by the Shanghai Meteorological Service. Six countries (Indonesia, Malaysia, the Philippines, Singapore, Thailand, Timor-Leste) provide full services (Category 3); one country, Vietnam, falls between Categories 2 and 3; two countries (the Lao People's Democratic Republic and Mongolia) have the capacity to provide essential services (Category 2); and three countries (Cambodia, Republic of the Union of Myanmar, Papua New Guinea) in addition to PICs provide basic services or less (Category 1 or lower). The most advanced NMSs provide effective MHEWSs. See box 5.2 and appendixes F and G for further information about individual categories.

Understaffing is one of the most serious problems followed by sustainable investment in infrastructure. The weakest NMSs and NHSs share problems, notably that their governments generally overlook their importance in disaster mitigation and economic development and severely underfund their operational costs. This causes limited observing networks and often obsolete or broken equipment, no calibration and maintenance facilities, limited forecasting tools, and poor service delivery. Coordination across government is also weak, resulting in limited capacity to respond to hazard warnings. The lack of staff is especially acute in the PICs, where qualified personnel migrate to the Australia Bureau of Meteorology or MetService in New Zealand.

Box 5.2 Weather and Climate Services Progress Model

The following composite criteria are adapted from WMO climate services and public weather services and expert opinion, which includes the capacity of the NMS or NHS to maintain an observing network and provide forecasts and climate services and deliver weather, water, and climate services to users.

1. Observing and Forecasting Systems
   Category 1: Basic Observations and Forecasting
   Category 2: Essential Observations and Forecasting
   Category 3: Full Observations and Forecasting
   Category 4: Advanced Observations and Forecasting
2. Weather Services Delivery
   Category 0: No Service Delivery
   Category 1: Basic Service Delivery
   Category 2: Essential Service Delivery
   Category 3: Full Service Delivery
   Category 4: Advanced Service Delivery
3. Climate Services
   Category 1: Basic Climate Services
   Category 2: Essential Climate Services
   Category 3: Full Climate Services
   Category 4: Advanced Climate Services

Pacific Island Countries (PICs)

Producing timely and accurate, and cost-effective, early warnings of weather-related hazards is a major challenge for many PICs, which have severe staffing and funding constraints. PICs have a relatively common structure and have historically depended on Australia and New Zealand (excluding the French territories) for training and operational support. Although conflict has influenced its effectiveness, the most capable NMS of the PICs is the Fiji Meteorological Service, which operates the WMO Regional Specialized Meteorological Center (RSMC) Nadi-Tropical Cyclone Center. This center is responsible for issuing tropical cyclone advisories for the Southwest Pacific Ocean, backed up by an extensive network of RSMCs and Tropical Cyclone Warning Centers (TCWCs; table 5.1). Small island states such as Kiribati, Samoa, and Tonga are dependent on the Fiji Meteorological Service for many basic services, such as aviation forecasts and warnings, forecasts for shipping, as well as general public forecasts. These services have been provided for free to these countries for many decades; however, increasing demands for services within Fiji are placing pressure on their future existence.

Regional cooperation in the Pacific is relatively strong. The current strategy of the Secretariat of the Pacific Regional Environmental Programme aims to enhance and build capacity for applied research, foster meteorological and oceanic observation and monitoring programs to improve understanding, and develop targeted responses to climate change and related disaster risk reduction. The goal is to provide adequate regional meteorological and oceanographic services to ensure access to quality and timely weather ocean state information. The target is that by 2015, at least 14 NMSs have improved access to tools and applied scientific knowledge of Pacific climate drivers and projections and have installed and implemented national climate and disaster databases (SPREP 2011).

Table 5.1 Tropical Cyclone Warning Centers in East Asia and the Pacific

Regional Specialized Meteorological Centers	
RSMC Tokyo Typhoon Center/Japan Meteorological Agency	Western North Pacific Ocean and South China Sea
RSMC Nadi-Tropical Cyclone Center/Fiji Meteorological Service	Southwest Pacific Ocean
Tropical Cyclone Warning Centers	
TCWC Perth/Bureau of Meteorology, Australia	Southeast Indian Ocean
TCWC Darwin/Bureau of Meteorology, Australia	Arafura Sea and Gulf of Carpentaria
TCWC Brisbane/Bureau of Meteorology, Australia	Coral Sea
TCWC Port Moresby/National Weather Service, Papua New Guinea	Solomon Sea and Gulf of Papua

Source: Authors.

Cambodia and Lao PDR

A closer look at Cambodia and Lao PDR highlights the different capacities of countries in the region that are trying to cope with common meteorological and hydrological hazards. Cambodia and Lao PDR are located in the heavy rainfall area of the Indochina Peninsula. Floods, droughts, and extreme weather affect both countries, which are highly dependent on rainfall for agriculture. Most flooding in the region occurs from May to September when monsoon rains accumulate in the upper Mekong River basin. In Lao PDR, the slopes of the mountains represent an important factor in draining heavy rains down rapidly from upstream through the rivers into the low-lying areas and causing serious flooding during this season. Some typhoons and tropical depressions that reach Indochina do not weaken over the land and produce torrential rainfall and extensive flooding, but their characteristics are not well understood. Most big cities are located in flat plains along the Mekong River and its tributaries, which are home to 50 percent of the total population. Thus they are vulnerable to disasters, which have significant socioeconomic impacts in terms of loss of lives and damage to properties' basic infrastructure, including transportation services. In Cambodia, normal flooding is an integral part of the agricultural system, which provides nutrients to the floodplains where much of the agricultural production is located. A system of canals and levees are used to manage the flow of water.

Both Cambodia and Lao PDR need to make substantial investments in the observational infrastructure, forecasting, and service delivery. Two case studies are presented (see boxes 5.3 and 5.4), based on presentations of the Department of Meteorology (DOM), Cambodia, and the Department of Meteorology and

---

Box 5.3 Case Study: Lao PDR

Located in the Lao PDR Ministry of Natural Resource and Environment, the Department of Meteorology and Hydrology (DMH) has a staff of 205 with 70 people at the DMH headquarters and 135 in the provinces. Fourteen staff supports the weather forecast office and aeronautical division. Other divisions include the Hydrological Division, Meteorological Network and Earthquake Division, and Climate Division. The DMH has 17 main and 32 secondary synoptic stations and 113 rain gauges. The secondary stations record data monthly. The DMH also operates C-band radar. The hydrological network consists of 109 staff gauges and 49 discharge stations. In addition, data are also received from the Mekong River Commission (MRC). Weather forecasting is based on six-hourly NOGAPS NWP products from the Fleet Numerical Meteorology and Oceanography Center, information from the World Area Forecast Center in the United Kingdom, and RSMC Tokyo. Additional data are also obtained from Vietnam. The Weather Forecast Office provides daily, three-day, and long-range (seven-day) weather forecasts as well as monthly and seasonal climate outlooks. River flood forecasting is limited to the main stream of the Mekong. MRC is developing a flash flood guidance system.

During hazardous weather, bulletins are issued two or three times a day. The NOGAPS model provides accumulated rainfall estimates, which can exceed 100 millimeters in 12 hours. Flash floods are frequent. The estimated cost of a single event in 2006 exceeded US$3 million in agriculture, livestock, irrigation, and infrastructure losses. Tropical cyclone damage can be much higher. In 2009 Cyclone Ketsana caused more than US$58 million in damage, including houses, schools, agriculture, and irrigation. More than 270,000 people were affected with

box continues next page

Box 5.3 Case Study: Lao PDR *(continued)*

28 deaths and nearly 100 injured. In 2011 Typhoons Haima and Nokten affected more than 1,300 villages and caused in excess of US$174 million in damages.

Lao PDR has developed a disaster management framework, which links the DMH with the National Disaster Management Office and through this with the provincial, district, and village disaster management committees. Radio is the most powerful tool for public awareness of weather, flood forecasts, and warnings. Television has less complete network coverage in the country. Internet services are available in most urban areas. People living in remote areas obtain weather and flood forecasts and warning information via the radio network, or they receive warnings and announcements through local authorities.

A new early warning strategy for flood forecasting has been developed in close consultation with DMH, National Disaster Management Offices, key development partners, and civil society organizations from all levels of the community. The specific role of MoNRE is to set up an early warning system for forecasting floods, drought, and other natural hazards, improve and upgrade the national hydromet network and earthquake monitoring stations, and manage, disseminate, and provide data and information on natural resources and the environment, including disasters and climate conditions and analysis, to the public. A legal framework has been established for National Disaster Management, which mandates the establishment of a MHEWS with DMH responsible for the technical component of the warning system.

Current gaps in the system include the need to upgrade and extend the observing networks and data communication systems, insufficient qualified staff, better coordination of data sharing, the need to increase numerical prediction capability and skills of forecasters to make use of new techniques, the need to translate technical guidance into information that can be used by people making decisions at the local level, and the need for more trainers and communicators at local levels to facilitate public awareness.

Source: Authors based on recent (2011) presentations of the Department of Meteorology and Hydrology (DMH), Lao PDR, and early reviews conducted on behalf of WMO, UNISDR, and the World Bank 2012.

Box 5.4 Case Study: Cambodia

Cambodia's Ministry of Water Resources and Meteorology is home to both the Department of Hydrology and River Works (DHRW) and the Department of Meteorology (DOM). The minister is the Permanent Representative of Cambodia with the WMO. Ninety percent of Cambodia's rainfall occurs during the southwest monsoon from May to November. Annual rainfall of 3,000–5,000 millimeters is recorded in the west of the country and 1,300–1,500 millimeters in the east. Maximum rainfall can exceed 200 millimeters in 24 hours because of intense convection, and occasionally typhoons bring strong winds and torrential rain and cause extensive damage and disruption. The principal causes of natural disasters in Cambodia are flooding, drought, epidemics, and storms.

box continues next page

Box 5.4 Case Study: Cambodia *(continued)*

Hydrological observations along the Mekong River and its tributaries have been sporadic and mostly project-based, with the majority supported by the Mekong River Commission. Staffing capacity is limited, making it difficult to maintain equipment and operate the network effectively from the DHRW in Phnom Penh. DOM employs a total of 44 staff. It operates 21 manual synoptic stations, eight automatic weather stations, and 200 manual rain gauges. There are no upper air stations, marine observations, or radar. The forecast office utilizes products available on the Global Telecommunication System producing 24-, 36-, and 72-hour forecasts using the NOGAPS model of the Fleet Numerical Meteorology and Oceanography Center in Monterey, California. They also use Ensemble Prediction System Meteograms produced by the Republic of Korea Meteorological Administration and receive support from RSMC Tokyo and the Hong Kong SAR, China, Observatory for typhoon warnings. The DHRW and DOM are responsible for flood and weather forecasts and warnings. Authorization to issue warnings is complex, requiring a series of approvals well above the forecasters responsible for the production of the warning information. This may limit the timeliness and effectiveness of the information, which is also not fully integrated into a disaster management system. Despite being in the same ministry, cooperation between the DHRW and DOM is limited.

DOM's own strategy is to strengthen and modernize its surface meteorological and observing and data communication networks, introduce Doppler radar to improve short-range flood and weather forecasting, and introduce local area modeling using numerical weather prediction tools developed for the region. Institutional strengthening of the DHRW and DOM is a priority to enable them to develop a comprehensive multihazard weather, climate, and water warning system. Stronger institutions will be able to support an enhanced hydrological observing network within the DHRW and meteorological observing network within the DOM. However, warning services go well beyond observations and must include hazard forecasting and the effective delivery of services. Flood forecasting on all time scales from nowcasts of flash floods to seasonal flooding depends on the integration of weather, climate, and hydrological information. This is a challenge for organizations that do not work closely together. Building the skill to provide early alerts and warnings is essential. Experience suggests that this can only be achieved through greater integration and cooperation between meteorologists and hydrologists. Such an opportunity exists within the Ministry of Water Resources and Meteorology.

Source: Authors based on recent (2011) presentations of the Department of Meteorology (DOM), Cambodia, and early reviews conducted on behalf of WMO 2011, UNISDR, and the World Bank 2012.

Hydrology (DMH), Lao PDR, and early reviews conducted on behalf of WMO, United Nations International Strategy for Disaster Reduction (UNISDR), and the World Bank (See DMH and MoNRE 2012; WMO 2011; World Bank 2012). The focus by Lao PDR on disaster management and the creating of a legal framework is an important step in creating a Multi-Hazard Early Warning System and identifying clearly the roles and responsibilities of each of the actors, including the DMH. In contrast, Cambodia lags behind its neighbor in terms of forecasting capability and the capacity of its National Committee for Disaster Management and civil society to require and make effective use of early warnings. DOM is particularly limited in human resources compared with other NMSs in the region. However, the capacity to take immediate advantage of such an investment appears much higher in Lao PDR than in Cambodia.

Where Do We Want to Be?

Increase the capacity of NMSs and NHSs to respond to societal needs to reduce risks from natural hazards and facilitate economic development. All NMSs and NHSs providing less than advanced services (Category 4) need strengthening with major emphasis on Category 2 and lower services, especially to improve their meteorological and hydrological hazard warnings. Papua New Guinea and the PICs in Region V and the Republic of the Union of Myanmar and the Indochina Peninsula countries (Cambodia, Lao PDR, Vietnam) in Region II are in need of major improvements to their services through national investment and stronger regional cooperation. Strategies exist in many countries of the region to improve their NMSs and NHSs to cope more effectively with meteorological and hydrological hazards. Implementation plans and instruments are required to improve the capacity to provide early alerts and early warnings of extreme events to enable people to avoid harm and to protect livelihoods and property. In particular, the goal should be to close the gap between the most and least capable NMSs and NHSs by strengthening the weakest in partnership with the most advanced services in the region.

Invest in NHSs and NMSs. In a modern NMS, forecasters will spend most of their time on 0–6 hour nowcasting, with longer-period predictions relying on objective model-based forecasts that utilize the large-scale, multimodel ensembles constructed by the major prediction centers (Mass 2012). Forecasters will increasingly issue hourly nowcasts of the current weather situation and how the weather system will evolve over a few hours. Incorporated into a MHEWS, this will strengthen the connection between NMSs, NHSs, and disaster management and civil protection and can exploit effectively the latest technologies to translate and communicate information about the impact of severe weather events to people in ways they understand. Best practices for MHEWS have been developed by several countries, most notably in the region by the CMA through the Shanghai Meteorological Service (Tang et al. 2012), providing the capacity to alert, warn, and respond to a wide range of meteorological, hydrological, and environmental hazards.

Reduce Risks of Multiple Hazards. Growing urbanization increases risks of multiple hazards often triggered by meteorological events, and these need to be managed through a combination of structural and nonstructural measures. The nonstructural approaches focus on multihazard warning and effective preparation and response, which demands ever-greater cooperation between NMSs, NHSs, disaster management committees and other government agencies, and civil society. Good weather forecasting is a prerequisite for effective end-to-end early alert and early warning systems.

Make long-term technical investments. Because flooding is the predominant hazard in the region, most NMSs need access to tools such as Doppler radar, which has proven to be the most valuable in detection of high-impact weather. Assimilating radar and satellite data into skillful, high-resolution local area models provides the forecaster with the ability to warn-on-forecast rather than warn-on-detection. This approach will provide longer lead times for severe weather forecasts from the current realizable limit of about 20 minutes for warn-on-detection approaches, thereby providing emergency managers with much early warnings of hazardous weather and more time to make effective decisions. Many NMSs in the region operate Doppler radar networks. Therefore, countries with little experience in this technology could benefit from more advanced NMSs in the region to maintain these systems while sustainable technical skills are acquired. These investments need to be long-term, 10–15 years, to allow sufficient time for the development of national skills, while ensuring the continuity of the observing systems.

Manage food and water security. Ensuring food security among the largely agrarian societies of the region is a priority. The capacity to cope with the vagaries of the monsoons, climate variability, and change is a must. This depends on accurate rainfall prediction on long time scales (months to seasons and beyond) to select appropriate crop varieties and manage water resources, and on shorter time scales to manage application of pest controls, fertilizers, and harvesting, and on the shortest scales to mitigate flash flood impacts on irrigation systems and other agricultural infrastructure. Working with water and agricultural sectors, NMSs and NHSs will provide sets of tools and services to harness the benefits of weather, climate, and water forecasts, as well as mitigating the adverse consequences of meteorological and hydrological hazards.

Promote regional cooperation. Although each PIC should have the national capacity to deliver weather and climate forecasts and warnings, it has been recognized that a regional approach to weather and climate services may be more sustainable in this region. Recently a broader strategy aimed at the sustainability of meteorological services has been developed (SPREP 2012). The strategy focuses on the primary hazards that have caused the largest losses of life and livelihood in the PICs, namely, tropical cyclones and typhoons, drought and flash floods, storm surges, earthquakes, and tsunamis (box 5.5). Australia, Finland, Japan, New Zealand, the United States, and the WMO have supported the development of this strategy. Also in place is the Pacific Meteorological Council (PMC),[6] which was formed in 2011 to lead efforts on regional coordination of and resource mobilization for National Meteorological Services and to strengthen cooperation between NMSs and NHSs and the Disaster Management Community.

Focus on targeted capacity building. Much capacity development and training has been undertaken to improve technical skills needed to deliver weather and climate services in PICs. Despite progress, much remains to be done to bring many NMSs up to the level that will ensure they can meet their mandates and serve their nations effectively. The current capacity varies greatly between NMSs. Most NMSs in the region operate with poor infrastructure and limited capability. Their climate services are generally poorly developed or nonexistent.

Box 5.5 Pacific Islands Meteorological Strategy 2012–21: Sustaining Weather and Climate Services in Pacific Island Countries and Territories

Vision: NMSs of the Pacific island countries and territories are able to provide relevant weather and climate services to their people to make informed decisions for their safety, socioeconomic well-being and prosperity, and sustainable livelihoods.

Key outcome: Improved early warning systems for floods (EWS-floods).

National priorities:

1. Establish and/or strengthen institutional capacity for EWS-floods.
2. Ensure that EWS-floods are integrated into government policies, decision-making processes, and emergency management systems at both national and community levels.
3. Complete inventories and needs analyses of national EWS-floods ensuring inputs from all stakeholders, including women, children, and disabled people, taking into full consideration traditional knowledge, and upgrade and/or redesign national EWS-floods to cater to these special needs.
4. Joint programs with National Disaster Management Offices, including conducting public awareness meetings, education, and analyses of socioeconomic impacts of floods and benefits of EWS-floods.
5. Strengthen relationships between NMSs and hydrological agencies that may have responsibility for issuing flood warnings.
6. Identify hydrological monitoring resources (for example, satellites) that may aid in flood warnings.
7. Support studies on socioeconomic benefits of EWS-floods.

Regional priorities:

1. In partnership with other agencies, assist NMSs in developing and strengthening EWS-floods, including the following:
   a. Coordinate regional support for implementing EWS-floods.
   b. Coordinate development of guidelines for EWS-floods.
   c. Identify and coordinate sharing of available tools or methodologies including the geographic information system, satellite information, and hazard mapping for EWS-floods.
2. Coordinate analyses of EWS-floods.

In a number of instances, PICs rely mainly on external support to provide basic climate services.

Improve forecasting skills in Cambodia and Lao PDR. Capacity building in these two countries is being undertaken through their participation in the WMO's Severe Weather Forecasting Demonstration Project (SWFDP) Southeast Asia, which aims to enable weather forecasters to access and effectively use numerical weather prediction (NWP) and ensemble prediction system[7] (EPS) products to improve forecasting of severe weather and related hazards, and to improve the utility of forecast-related services for users. Also involved in this activity are Thailand and Vietnam. The SWFDP provides each of these countries with access to NWP, EPS, and satellite products from a variety of WMO sources including the China Meteorological Administration (CMA), Japan Meteorological Agency (JMA), Republic of Korea Meteorological Administration (KMA), and the regional centers of the Hong Kong Observatory (HKO), China, RSMC Tokyo (typhoon forecasting), and RSMC New Delhi (tropical cyclone forecasting) and the European Centre for Medium-Range Weather Forecasts (ECMWF). Once completed, it is expected that the forecasting skill of each of the national meteorological centers will have improved along with their ability to translate probabilistic forecasts into information that can be used effectively by decision makers. The ongoing, operational success of this activity will depend on continuing access to products from the partner NMSs capable of generating and assembling the ensemble predictions for the region.

In East Asia and the Pacific there is already extensive regional cooperation on tropical cyclone forecasting through the WMO Regional Centers. Given the high capacity of many of NMSs in the region, this could extend to operational pairing arrangements that would improve access to routine as well as severe forecasts in the less developed NMSs. Supporting the implementation of the regional strategy for the PIC NMSs and NHSs may be a good starting point. Regional investment would include BOM, Australia, and MetService, New Zealand. Cooperation on the Indochina Peninsula could also be supported beyond the existing Southeast Asia SWFDP, which would directly benefit Cambodia, Lao PDR, and Vietnam. Pairing with CMA, JMA, and KMA is an option. The Republic of the Union of Myanmar would be part of the grouping currently developing the Bay of Bengal SWFDP and would be paired with the Indian Meteorological Department and the Thai Meteorological Department. Mongolia already benefits from close ties with neighbors, CMA and Roshydromet, which could be further strengthened. The remaining Category 3 services could also benefit from stronger regional cooperation, especially better data sharing, which would improve both weather and climate outlooks in the region. Achieving broader regional cooperation would require support to existing regional operational structures, such as the WMO Regional Climate Centers and Regional Meteorological Centers as part of future national and regional investments to improve meteorological and hydrological services.

What Needs to Be Done?

Modernization of NMSs and NHSs needs to consider how best to develop meteorological and hydrological services that effectively combine both national and regional investment. Although it is clear that basic infrastructure improvements are needed in many NMSs and NHSs, sustaining their forecasting and warning services is also likely to require routine operational support from more advanced NMSs. The success of SWFDPs in other regions indicates that the solution may be a combination of strengthening the national institutions and building capacity in the more advanced countries to provide regional support. National strengthening of NMSs and NHSs should focus on three elements:

- **Institutional strengthening, capacity building, and implementation support,** which includes strengthening the NMS's legal and regulatory framework; improving its institutional performance as the main provider of weather, climate, and hydrological information for the country; building the capacity of personnel and management; ensuring operability of future networks; and supporting project implementation.

- **Modernization of observation infrastructure and forecasting,** which includes modernizing the NMS's and NHS's observation networks, communication, and information and communication technology systems; improving the meteorological and hydrological forecasting system; and refurbishing NMS offices and facilities.

- **Enhancement of the service delivery system,** which includes creating or strengthening the public weather and climate service; and developing new information and value-added products for vulnerable communities and the main weather- and climate-dependent sectors of the country's economy, for example, by means of a functioning accreditation system for suppliers of hydromet equipment and investments in regional (or national) centers capable of supplying spare parts and maintenance services.

This cannot be done piecemeal but must be considered as a transforming process to raise the overall capability of the organization and improve the chances of sustaining the modernization effort. It also requires acceptance by government of the expected increase in the operation and maintenance costs associated with the modernization of the observing networks and forecasting systems. Proposed investments should be supported by awareness campaigns among national and government stakeholders explaining the importance of advanced hydrological and meteorological services for disaster reduction and economic development.

1. **The modernization should ensure that there is an effective legal framework for meteorological and hydrological services and clear definitions of responsibilities for warnings.** There should be standard procedures for information sharing, and sufficient base resources to support the operations and maintenance of the observing networks. Forecasters should have access to the latest forecasting techniques, and there should be an operational focus on delivering services that meet users' needs. Proposed investments should be supported by awareness campaigns among national and government stakeholders explaining the importance of advanced hydrological and meteorological services for disaster reduction and economic development.

2. **Greater cooperation and integration between NMSs and NHSs are essential.** Weather, water, and climate are not independent. The benefits of integrated monitoring and forecasting should be recognized and every effort made to reduce the gaps between these disciplines. The tendency to separate these along functional lines should be avoided.

3. **Introducing MHEWS,** based on the existing experience in the region, will benefit many weaker countries, establishing a discipline for early warning and greater communication between NMSs, NHSs, and disaster management agencies.

4. **Greater emphasis in modernization programs will be on nowcasting** (0–6 hours) of local weather events, relying on objective model-based guidance from major WMO prediction centers for one-day or longer outlooks. Nowcasts of high-impact weather, especially flash floods, depend on complex technologies, including radar, satellites, and local area numerical models. The challenge is how to sustain these technologies in the NMSs and NHSs that have limited technical capacity.

5. **Considerable scope exists for more effective regional cooperation beyond the existing coordination for tropical cyclone warnings.** Given the high capacity of many NMSs in the region, this could extend to operational pairing arrangements that would improve access to routine as well as severe forecasts in the less developed NMSs. Support to existing regional operational structures, such as the WMO Regional Climate Centers and Regional Meteorological Centers, is recommended as part of future national and regional investments to improve meteorological and hydrological services.

How Can the World Bank Help?

Realizing the large potential that hydromet services offer, the World Bank has significantly scaled up its efforts in the hydromet field in the last few decades. Since the mid-1980s, the Bank has prepared and implemented more than 130 operations for hydromet forecasting. The current total cost of investments under preparation or implementation exceeds US$500 million. The projects have actively supported national NMS and NHS through institutional strengthening and capacity building, modernization of observation networks, and improvement of service delivery. Recently the World Bank gained experience in modernization of weather and climate services to provide assistance to governments and close the gap between the most and least capable NMSs in the region by strengthening the weakest in partnership with the most advanced services in the region. In addition, the World Bank together with the Global Facility for Disaster Reduction and Recovery created the Weather and Climate Information and Decision Support Systems unit, which is a group of hydromet experts advising governments on best solutions in this area. In East Asia and the Pacific, the

World Bank has been advancing hydromet services on several fronts. Some examples are given below.

Providing technical assistance and investment lending. In several countries, the World Bank investments in regional NMSs and NHSs support institutional strengthening, improving observation networks and forecasting, and ensuring more effective service delivery, to transform weak, unsustainable institutions into ones with the capacity to meet national expectations for weather, water, and climate services.

Supporting integrative strategies. The World Bank's approach in Vietnam supports an integrated method, which combines improving weather-monitoring infrastructure, strengthening institutions, and delivering improved products on national, regional, and local levels. In the past, Vietnam has had several initiatives that targeted improvement of the hydromet monitoring infrastructure. Yet these efforts have been fragmented and lack an integrated vision and adequate coordination among relevant institutions. The World Bank, through the Vietnam Managing Natural Hazards Project, addresses this shortcoming through preparing and implementing an integrated hydromet framework, including a nationwide communication system and an end-to-end warning system. The project modernizes the forecasting system and develops the capacity for improved operations that will help to prepare high-quality forecast products and early warnings.

Investing in last-mile early warning. In Lao PDR, the World Bank, with Global Facility for Disaster Reduction and Recovery support, is addressing weaknesses in the dissemination of the Early Warning Bulletin to local communities. In the past, the Early Warning Bulletin often stopped at the province level because of lack of further communication mechanisms to the community. In a new project, reliable information and communication technologies are being used to strengthen last-mile warning dissemination mechanisms. Along with the installation of equipment, regulations are being drafted that stipulate emergency communications and community drills. Finally, provincial and district officials are trained in interpreting the messages contained in the bulletin and in acting upon them.

Strengthening legal frameworks. Legal frameworks have the important function to assign roles and responsibilities and to improve accountabilities. In Lao PDR, a decree is being drafted for regulations on prevention of and preparedness for meteorological disasters. The decree authorizes the responsible institutions and regulates hydromet services. It designates tasks related to early warning systems, development and maintenance of hydrological and meteorological observation, and international cooperation.

Supporting regional data sharing. The World Bank, jointly with the WMO and the UNISDR, is also exploring the opportunity of data sharing between countries in East Asia, particularly between the Mekong basin countries of Cambodia, Lao PDR, and Vietnam. The objectives of these efforts are to enhance coordination among hydromet warning systems and to strengthen regional harmonization and interoperability of the observing networks and data.

Promoting partnerships, enhancing knowledge transfer, and leveraging investments. An important element of the Pacific Islands Meteorological Strategy 2012–21 is the recognition that, where appropriate, services may be delivered by NMSs with greater resources in support of those with less. In some cases, depending on available resources, it may be more efficient to deliver certain services and support at a regional level, subject to bilateral and multilateral agreements. The strategy also recognizes that partnerships with the WMO, regional intergovernmental agencies and organizations, and technical partners are critical. A multilateral coordinated approach, which the World Bank helped to develop, enhances effectiveness in increasing resources and targeting efforts and managing potential overlap between agencies, organizations, and development partners, especially where these are managed through bilateral arrangements. Partnerships between NMSs have an important role in ensuring cooperation and sharing lessons learned within the region.

Notes

1. The benefits accrue to the public through avoided losses in day-to-day activities and to the economy in three ways: governmental decision making in organized disaster preparation and mitigation, economic efficiency through appropriate management, and specialized services to acquire benefits and avoid or reduce economic losses.

2. Also known as typhoons in the northwestern Pacific. Typhoon Tip in the northwestern Pacific on October 12, 1979, was measured to have a central pressure of 870 mb and estimated sustained winds of 305 km h^{-1} (190 miles per hour), making it the most intense tropical cyclone on record.

3. There is a well-defined interdecadal variation in tropical cyclone activity in the northwest Pacific. For example, 1998–2010 was relatively inactive compared with 1989–98. It has been proposed that this is related to strong vertical wind shear and strong subtropical high pressure in the region of tropical cyclone genesis.

4. The East Asia and Pacific region as defined by the World Bank includes 15 countries plus PICs and territories; from a meteorological and hydrological perspective, two WMO regions (II and V), which include 26 countries and 10 PICs and territories (see appendix F), overlap this region. These WMO regions contain some of the most advanced and least advanced NMSs and NHSs in the world. The combined capacity of Australia, China, France, India, Japan, Korea, New Zealand, Russia, the United Kingdom, and the United States make up most of the global investment in NMSs. For the purposes of this assessment, we have developed a composite classification for NMSs in the region, which has been adapted from separate classifications of WMO climate services and public weather services and the authors' opinion of observing and forecasting systems (see appendixes F and G).

5. For example, Météo France provides the weather services for New Caledonia and French Polynesia.

6. The PMC is supported by the WMO, U.S. NOAA National Weather Service, MetService of New Zealand, NIWA (National Institute of Water and Atmospheric Research), Australian Bureau of Meteorology, Météo France, JICA, and the Finland Ministry of Foreign Affairs and Finnish Meteorological Institute, plus bilateral and multilateral donors including Australia, the United States, and the World Bank.

7. Ensemble prediction systems (EPSs): Multiple numerical predictions are conducted using slightly different initial conditions that are all plausible given the current and past set of observations. The system may use a single model or multiple models. Each of the individual simulations creates a member of the ensemble, which together provides the forecaster with information on the uncertainty of the forecast. This system allows the forecaster to identify high-impact, low-probability events, which might otherwise go undetected.

References

Chan, J. C. L. 2008. "Decadal Variations of Intense Typhoon Occurrence in the Western North Pacific." *Proceedings of the Royal Society A* 464: 249–72.

Department of Meteorology and Hydrology (DMH) and Ministry of Natural Resources and Environment (MoNRE). 2012. "Early Warning Strategy for Flood in Lao PDR."

Hallegatte, S. 2012. "A Cost-Effective Solution to Reduce Disaster Losses in Developing Countries: Hydro-Meteorological Services, Early Warning, and Evacuation." Policy Research Working Paper 6058, World Bank, Washington, DC.

Jarraud, M. 2007. WMO Secretary-General, Opening Statement of Madrid Conference "Social and Economic Benefits of Weather-, Climate- and Water-Related Information and Services." Madrid, March 19–22.

Lazo, J. K., R. E. Morss, and J. L. Demuth. 2009. "300 Billion Served." *Bulletin of the American Meteorological Society* 90: 785–98. http://www.isse.ucar.edu/staff/rebeccam/pdf/LazoMorssDemuth_earlyonline.pdf.

Mass, C. 2012. "Nowcasting: The Promise of New Technologies of Communication, Modeling and Observation." *Bulletin of the American Meteorology Society* 93: 797–809. http://www.atmos.washington.edu/cliff/NewFInal010912.pdf.

SPREP (Secretariat Pacific Regional Environment Programme). 2011. *Pacific Regional Environment Programme Strategic Plan 2011–2015.* Apia: SPREP.

———. 2012. *Pacific Islands Meteorological Strategy 2012–2021: Sustaining Weather and Climate Services in Pacific Island Countries and Territories.* Apia.

Tang, X., L. Feng, Y. Zou, and H. Mu. 2012. "The Shanghai Multi-Hazard Warning System: Addressing the Challenge of Disaster Risk Reduction in an Urban Megalopolis." In *Institutional Partnerships in Multi-Hazard Early Warning Systems*, edited by M. Golnaraghi, 159–79. New York: Springer.

WMO (World Meteorological Organization). 2011. *Regional Subproject Management Team (RSMT) of the Severe Weather Forecasting Demonstration Project (SWFDP) in Southeast Asia.* Commission for Basic Systems, Ha Noi, October 10–13.

World Bank. 2012. *Review of Hydrological and Meteorological Activities in Cambodia.* Washington, DC.

Zhang, G., and W. Haixiao. 2003. "Evaluating the Benefits of Meteorological Services in China." *WMO Bulletin* 52: 383–87.

CHAPTER 6

Financial Protection: Risk Financing and Transfer Mechanisms

Key Messages for Policy Makers

- Disasters inflict extreme financial and fiscal tolls across countries in East Asia and the Pacific, with governments shouldering an increasing financial responsibility for post-disaster recovery and reconstruction. Most regional governments have inadequate and/or incomplete funding arrangements in place for major disasters, which can significantly exacerbate the adverse socioeconomic consequences of these events.
- National disaster risk-financing strategies should aim to increase governments' post-disaster response capacity while protecting their long-term fiscal balance.
- Disaster risks should be integrated into fiscal risk and public debt management strategies. Post-disaster budget allocation and execution processes should ensure effective and timely post-disaster response.
- Governments should aim to reduce their contingent liability to disasters by integrating disaster risk considerations into investment frameworks for public assets.
- Governments should leverage the financial and technical capacity of the private insurance and reinsurance markets and support the development of competitive domestic catastrophe risk insurance markets.

Where Are We Now?

Recent disasters provide compelling reminders of the extreme financial and fiscal tolls that these events inflict. In Thailand the 2011 floods resulted in approximately US$46.5 billion of damage and losses and required government spending amounting to 5 percent of the government's annual revenues (World Bank 2012b). In the Solomon Islands, the 8.1 magnitude earthquake followed by a tsunami that hit in April 2007 caused losses estimated at 95 percent of the government's budget and created a short-term liquidity crunch until donor assistance arrived. More recently,

---

This chapter was written by Olivier Mahul and Laura Boudreau.

the tsunami that hit Samoa in September 2009 caused losses estimated at 22 percent of national gross domestic product (GDP). These examples illustrate a broad trend of increasing catastrophic losses from disasters occurring across diverse countries in East Asia and the Pacific, with governments shouldering an increasing financial responsibility for post-disaster recovery and reconstruction.

The World Bank has developed a disaster risk-financing and insurance (DRFI) framework for understanding and improving the financial resilience of states against disasters. The DRFI framework promotes a dual approach to increasing overall financial resilience based on financial disaster risk assessment and modeling (figure 6.1, see also Cummins and Mahul 2010). This approach includes the following: (1) sovereign disaster risk financing, which entails identification and assessment of the government's contingent liabilities associated with natural hazards and financial strategies to increase their financial response capacity in the aftermath of a disaster while protecting their long-term fiscal balance, and (2) catastrophe risk market development, which increases the transfer of public and private risks to the insurance sector.

Engagement in disaster risk financing and insurance in East Asia and the Pacific is variable because of the diverse mix of countries in the region, particularly China. The financial and fiscal impacts of disasters are extremely different, for example, in a Pacific island state with limited geographic and economic diversification (where a disaster can inflict losses several times GDP) versus a large country with a diversified economic base, for example, China or Indonesia. Although this is the case, because of the high hazard and exposure levels in the region, all countries in East Asia and the Pacific can benefit from engagement in DRFI that is adapted to their specific needs and situation.[1]

Figure 6.1 Increasing Society's Financial Resilience to Disasters

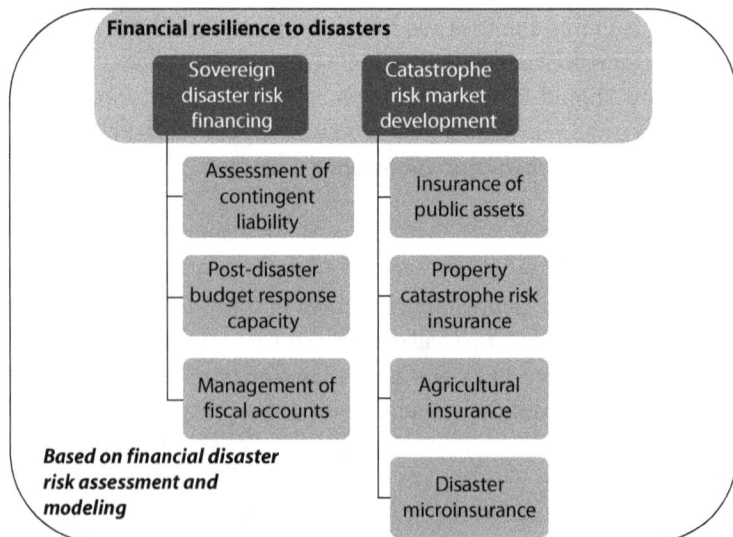

Source: World Bank Disaster Risk Financing and Insurance Program 2012.

Financial Risk Assessment in East Asia and the Pacific: What Do We Know?

Financial resilience to disasters in East Asia and the Pacific starts with understanding the financial and fiscal risks posed by these events. Analysis of historical disaster losses and probabilistic catastrophe risk modeling provides catastrophe risk metrics (such as annual expected loss and probable maximum losses), which, combined with financial and actuarial tools, can guide financial decision making.

Availability of catastrophe risk assessment for financial applications is variable across countries in East Asia and the Pacific. Primarily because of limited historical disaster loss records and low catastrophe risk insurance penetration, financial disaster risk information tends to be limited in countries in the region. In some countries with more developed insurance markets, proprietary catastrophe risk models have been developed by model vendors and insurance market participants. These models, however, cover only certain perils and may be outdated. Figure 6.2 maps catastrophe risk model availability for three major perils in East Asia and the Pacific from the three most prominent catastrophe risk-modeling firms: Risk Management Solutions (RMS), AIR Worldwide, and EQECAT.

The World Bank, in partnership with regional entities and national governments, has conducted financial risk assessment for several countries in East Asia and the Pacific. One of the most advanced financial disaster risk assessment programs has been conducted in 15 Pacific island countries (PICs) under the Pacific Catastrophe Risk Assessment and Financing Initiative. This project generated the most comprehensive collection of geospatial information for PICs as well as country-specific catastrophe risk models for earthquakes and cyclones (box 6.1). Preliminary financial catastrophe risk profiles have been prepared for ASEAN member states,[2] including more in-depth work in Indonesia, the Philippines, and Vietnam (box 6.2).

Findings of World Bank Financial Disaster Risk Assessments[3]

Pacific island countries: Each year, on average, PICs experience damage caused by natural disasters estimated at US$284 million, or 1.7 percent of regional GDP. As a percentage of GDP, Vanuatu and Tonga experience the largest annual average disaster losses, with 6.6 and 4.4 percent, respectively. Once every 75 years, it is expected that losses will exceed US$1.3 billion, or 7.8 percent of regional GDP. In this case, as a percentage of GDP, Tonga and Vanuatu are most affected, with greater than 50 and 40 percent of GDP lost, respectively (figure 6.3).

ASEAN member states: Each year, on average, ASEAN countries suffer damage in excess of US$4.4 billion as a consequence of natural hazards—equivalent to greater than 0.2 percent of regional GDP. Cambodia, the Lao People's Democratic Republic, Republic of the Union of Myanmar, the Philippines, and Vietnam face particularly high annual average expected losses relative to the size of their economies, equivalent to 0.7 percent or more of GDP. Every 100 years, on average, it is expected that losses will exceed US$17.9 billion, or an estimated 1 percent of regional GDP. Lao PDR and Cambodia face the highest expected losses relative to GDP for a 1-in-100 year event, at 11.7 and 7.3 percent, respectively (figure 6.4).

Figure 6.2 Catastrophe Model Vendor Coverage of East Asia and the Pacific

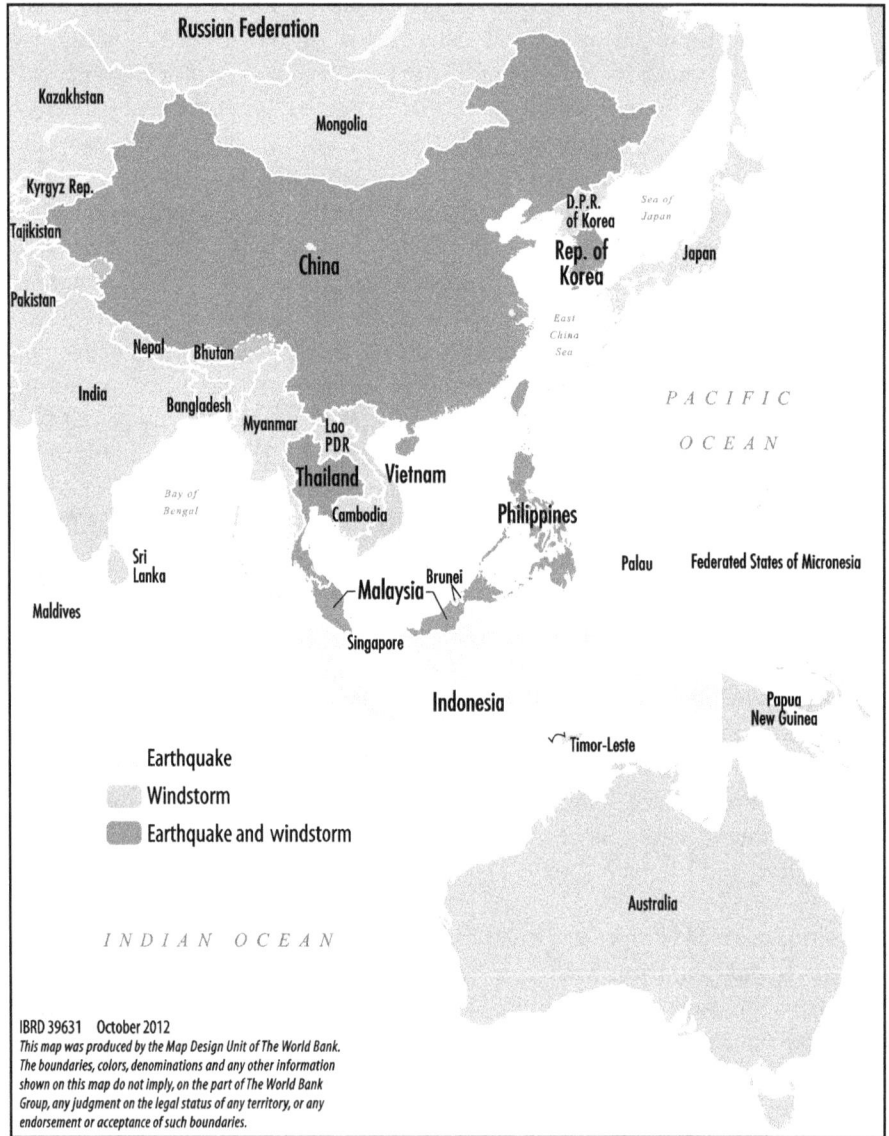

China: The World Bank has not conducted a detailed financial risk assessment for China. However, based on 10 years of historical loss data from Swiss Re (2002–11) (Swiss Re 2012a, 2012b), historical loss metrics for direct damages can be approximated. On average, China incurs US$27.4 billion, approximately 0.4 percent of GDP, of damages each year, with significant variability across years (figure 6.5). The highest loss year, 2008, was primarily driven by the highest loss event during this period, the Sichuan earthquake. In 2008 losses

Box 6.1 The Pacific Catastrophe Risk Assessment and Financing Initiative

The Pacific Catastrophe Risk Assessment and Financing Initiative (PCRAFI) provides PICs with a disaster and climate risk information system and associated tools for enhanced risk management to inform development planning and financing decisions. PCRAFI produced the Pacific Risk Information System, the most comprehensive system of geospatial information in the Pacific, which contains detailed, country-specific information on assets, population, and hazards. The platform can be used for many different risk management applications, for example, disaster risk financing and insurance. The initiative aims to engage in a dialogue with PICs on integrated financial solutions for the reduction of their financial vulnerability to disasters and to climate change.

In January 2012 PCRAFI entered its third phase, with one specific DRFI application: the Pacific Disaster Risk Financing and Insurance Program. This program provides institutional capacity building on disaster risk financing and insurance for the PICs as well as pilot sovereign parametric catastrophe risk insurance. The pilot, planning to launch in the fall of 2012 with five PICs (the Marshall Islands, Samoa, the Solomon Islands, Tonga, and Vanuatu), will be the first-ever transfer of catastrophe risk from the Pacific to the international reinsurance markets. Implemented over two years, the pilot will test the feasibility of a catastrophe risk pooling and transfer approach for the Pacific.

PCRAFI is a joint initiative between the Secretariat of the Pacific Community SOPAC, the World Bank, and the Asian Development Bank, with financial support from the government of Japan, the Global Facility for Disaster Reduction and Recovery (GFDRR), and the European Union, and technical support from Air Worldwide and the New Zealand GNS Science. The PICs involved in PCRAFI are the Cook Islands, Fiji, Kiribati, the Marshall Islands, the Federated States of Micronesia, Nauru, Niue, Palau, Papua New Guinea, Samoa, the Solomon Islands, Timor-Leste, Tonga, Tuvalu, and Vanuatu.

Box 6.2 National-Level Financial Catastrophe Risk Profiling

The World Bank has supported country-level financial and fiscal catastrophe risk assessment in three countries in East Asia and the Pacific: Indonesia, the Philippines, and Vietnam. In Indonesia and Vietnam, the assessments were preliminary profiles developed to provide the governments with the order of magnitude of possible losses to the public sector from disasters. In the Philippines, a more detailed catastrophe risk assessment relying on probabilistic catastrophe risk modeling is under way to be used to design and implement a disaster risk-financing strategy.

In Indonesia, a fiscal risk profile was constructed for the government based on historical disaster damage data and the estimated fiscal costs of destroyed buildings. The fiscal risk profile suggested that annual fiscal losses from disasters are in the range of US$420–US$500 million

box continues next page

Box 6.2 National-Level Financial Catastrophe Risk Profiling *(continued)*

and that once every 10 years this range of losses nearly doubles to US$800–US$950 million. The report also reviewed the legal and institutional framework for financial management of disasters and the execution of funds following a disaster. It also considers catastrophe insurance of public and private assets; interestingly, it found that although the central government is legally prohibited from purchasing catastrophe insurance, this practice was occurring at the subnational level.

In the Philippines, catastrophe risk assessment is being advanced. This approach relies on probabilistic catastrophe risk modeling of earthquakes, floods, and tropical cyclones in the country. It will enable quantification of possible financial losses from these events due to damage to government assets and contingent liabilities (for example, schools, hospitals, public buildings, large-scale infrastructure, and roads). The assessment will also include an inventory of the population at risk, enabling modeling of the potential numbers of displaced persons, injuries, and fatalities. The government can use these outputs to report on this risk in its annual fiscal risk statement and to design a risk-financing strategy, including risk transfer. In combination with additional information on the Philippines' macroeconomic conditions and on reinsurance market conditions, the assessment outputs will allow for the design of possible parametric risk transfer products as part of the government's risk-financing strategy.

In Vietnam, the assessment included a financial risk assessment for the country based on historical disaster losses, a review of the government's budgetary process for financing disasters, and a dynamic government funding gap analysis. The dynamic government funding gap analysis compared estimated disaster response and reconstruction costs with estimated available short- and long-term government resources. The analysis identified that short-term resources from contingency budgets and other sources are generally adequate for short-term needs, but reconstruction funding gaps were identified in some previous years and could be more acute in future years.

Source: Authors, with information from World Bank 2010 and World Bank 2011.

totaled approximately 3.3 percent of GDP. Of the catastrophe losses incurred during this 10-year period, only 1.2 percent were insured. For the Sichuan earthquake, less than 0.3 percent of losses were insured.

Fiscal Risk Management of Disasters: How Equipped Are Countries in East Asia and the Pacific to Finance Expected Losses?

Public responsibilities in the event of disaster vary across countries in East Asia and the Pacific, resulting in different contingent liabilities of their governments. The government's contingent liability can be defined as either explicit (that is, mandated by law, such as restoration of government assets and services) or implicit (that is, imposed by social or political expectations, such as post-disaster

Figure 6.3 Expected Loss Metrics for PICs

% of GDP

a. Estimated average annual loss

b. Estimated 75-year and 250-year probable maximum loss

■ 250-year loss   ▨ 75-year loss

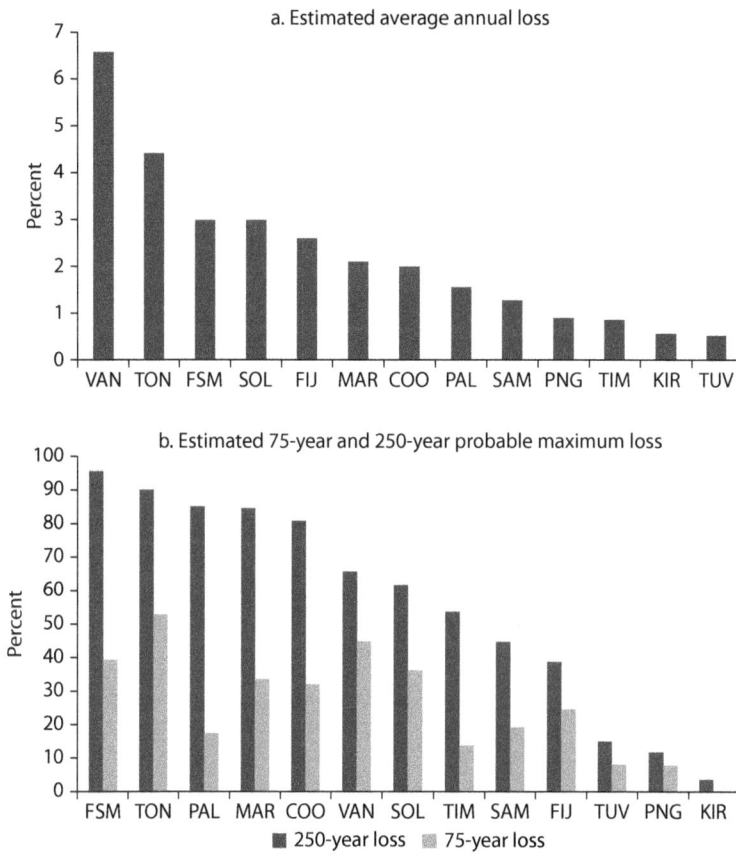

Source: Pacific Catastrophe Risk Assessment and Financing Initiative 2011.
Note: COO = Cook Islands; FIJ = Fiji; FSM = Federated States of Micronesia; KIR = Kiribati; MAR = Marshall Islands; PAL = Palau; PNG = Papua New Guinea; SAM = Samoa; SOL = Solomon Islands; TIM = Timor-Leste; TON = Tonga; TUV = Tuvalu; VAN = Vanuatu.

Figure 6.4 Expected Loss Metrics for ASEAN Member States

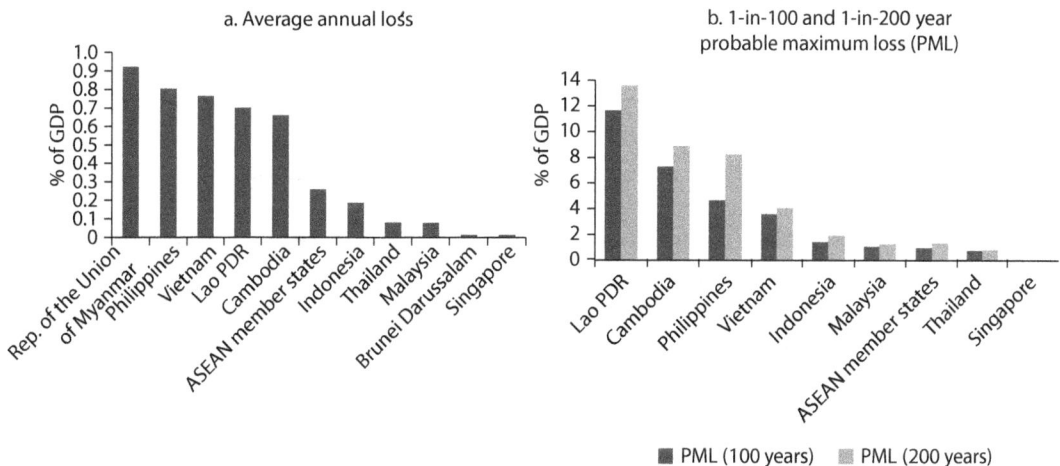

a. Average annual loss

b. 1-in-100 and 1-in-200 year probable maximum loss (PML)

■ PML (100 years)   ▨ PML (200 years)

Source: World Bank 2012a.
Note: Original data from EM-DAT CRED 2012 and Willis Research Network.

Figure 6.5 Disaster Losses for China, 2002–11

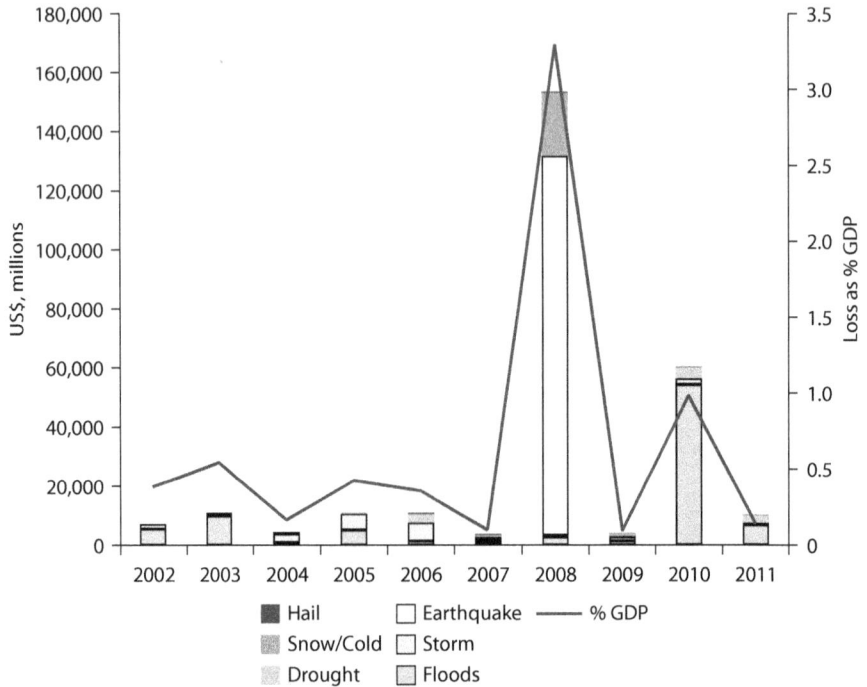

Sources: World Bank Disaster Risk Financing and Insurance Program, with data from Swiss Re 2012a, 2012b.

aid for restoration of private assets). Often it is difficult to identify the government's complete contingent liability in the case of a disaster because of vague legal wordings and variations in spending policies for historical events. In Vietnam, for example, a World Bank report found that provinces sometimes provide much higher levels of disaster compensation—in some cases five to six times more—than statutory levels under Vietnamese law (World Bank 2010).

Limited information is available on the contingent liability of governments in East Asia and the Pacific to disasters. The World Bank financial risk assessments of PICs and ASEAN member states contain information on expected fiscal costs from disasters. For the Pacific, emergency losses incurred by the government in the aftermath of a disaster (for example, debris removal) were estimated from the estimated direct (ground-up) losses (table 6.1). For the ASEAN region, the government's share in recovery and reconstruction spending was estimated based on observed ratios of damage to public sector recovery and reconstruction spending. These methods have limitations, because public sector spending on disasters can vary widely for a variety of reasons. Therefore, the estimates in this report should be taken as indicative and interpreted with caution.

The contingent liability estimates for PICs and ASEAN member states are indicative of the significant fiscal impacts of disasters in East Asia and the Pacific. For PICs, a 1-in-50 year earthquake, tsunami, or cyclone is estimated to

Table 6.1 Expected Emergency Losses from Disasters for PIC Governments

% of total government expenditures

PIC	AAL[a]	50 year	100 year
Fiji	2.5	19.2	26.3
Micronesia, Fed. Sts.	1.2	10.8	22.0
Kiribati	0.1	0.1	0.7
Marshall Islands	0.7	7.5	14.9
Nauru	0.0	0.0	0.0
Niue	1.3	13.3	31.6
Palau	0.8	5.3	13.4
Cook Islands	1.4	16.8	30.3
Papua New Guinea	1.0	11.2	15.7
Samoa	1.0	11.2	15.7
Solomon Islands	1.3	11.6	16.4
Timor-Leste	0.2	1.6	3.9
Tonga	3.2	28.3	42.1
Tuvalu	0.1	1.5	2.6
Vanuatu	5.7	34.3	43.6

Source: World Bank 2011.
a. AAL = average annual loss.

inflict emergency response costs for the government totaling at least 10 percent, and much greater in many countries, of the budget in 9 of 15 PICs—and these figures do not even take into account much more significant reconstruction costs. In ASEAN member states, three governments spend at least 1 percent of their budget on recovery and reconstruction costs each year, with the potential for much greater fiscal impacts from more extreme events (figure 6.6).

Several governments in East Asia and the Pacific have insufficient funding arrangements in place for major disasters, which can significantly exacerbate their adverse socioeconomic consequences. Available information suggests that governments in East Asia and the Pacific heavily rely on ex post instruments (for example, budget reallocation, debt, donor assistance) to finance the costs of disasters. Overreliance on ex post instruments entails significant fiscal uncertainty, because these require time to mobilize and can be costly in a post-disaster environment. Thus, the World Bank DRFI Framework promotes complementing ex post instruments with ex ante budgetary and, eventually, market-based instruments. Ex ante instruments (such as reserve funds, contingent credit, and insurance) provide immediate liquidity in the aftermath of a disaster, allowing the government to enact a swift and efficient response.

Several governments in East Asia and the Pacific are taking a proactive stance toward potential catastrophes by moving to an ex ante approach to disaster risk financing, as part of their broader disaster risk management (DRM) and climate change adaptation agenda. In recent years governments across the region have demonstrated interest in building their capacity and implementing financial instruments as part of the development of a disaster risk-financing strategy.

Figure 6.6 Expected Recovery and Reconstruction Liability of ASEAN Governments
% of total government expenditures

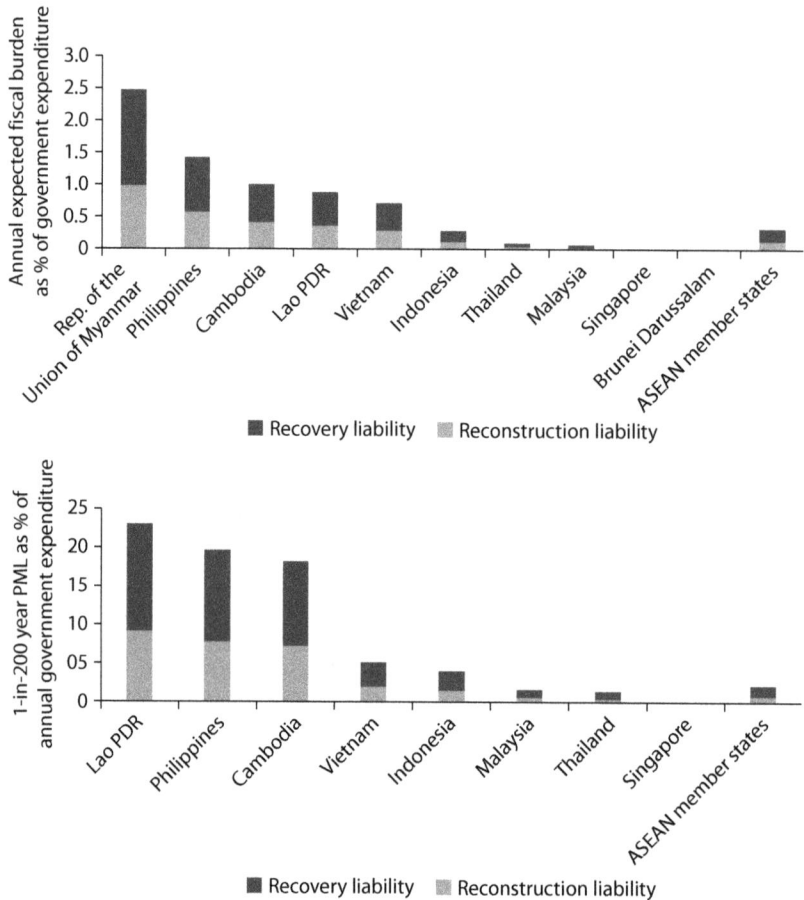

Recovery liability   Reconstruction liability

Recovery liability   Reconstruction liability

Source: World Bank 2011.
Note: Singapore and Brunei Darussalam did not present sufficient number of loss years, either historically or simulated, to compute reliable PMLs. Limited data were available for the Republic of the Union of Myanmar, and therefore its average annual loss may not accurately reflect long-term average annual losses. PML = probable maximum loss.

In 2011 the Philippines became the first country in East Asia and the Pacific to take advantage of a World Bank Catastrophe Deferred Drawdown Option (Cat-DDO) of US$500 million, a contingent credit product that provides immediate access to liquidity upon declaration of a state of emergency; the Philippines drew down the full amount of its Cat-DDO following Tropical Storm Washi to finance recovery and reconstruction costs. ASEAN member states more broadly have engaged in regional cooperation on the topic: Disaster risk financing was highlighted as an area for regional cooperation at the 2011 ASEAN+3 Finance Ministers' Meeting, and the countries agreed on a regional road map for engagement in DRFI at the regional and national levels (box 6.3). In the Pacific, five

Box 6.3 Toward Regional DRFI Cooperation by ASEAN Member States

Disaster risk financing and insurance was identified as an area for future regional financial cooperation at the ASEAN+3 Finance Ministers' Meeting in May 2011. ASEAN+3 Finance Ministers requested their deputies to initiate studies on DRFI. A taskforce on DRFI was established and has met twice since autumn 2011 to develop the agenda on DRFI in the region. Three main objectives have been identified for the region: (1) improving the perception and understanding of the economic and fiscal impacts of disasters on ASEAN member states; (2) improving institutional capacity to devise and implement cost-effective strategies for the fiscal protection of the state against disasters; and (3) promoting collaboration with international and capital markets, including a feasibility study on regional risk transfer mechanisms. With the prioritization of these three pillars, ASEAN+3 partners are moving toward a programmatic approach to DRFI for the region.

In November 2011 ASEAN countries, the World Bank, Global Facility for Disaster Reduction and Recovery, and the United Nations Office for Disaster Risk Reduction (UNISDR) convened the DRFI Forum in Jakarta, Indonesia. The regional DRFI initiative formed part of a broader program to strengthen disaster risk reduction capacity in the region, the focus of a Memorandum of Cooperation signed by the World Bank, ASEAN Secretariat, and UNISDR in 2009. DRFI was identified as a flagship program under the work plan for the ASEAN Agreement on Disaster Management and Emergency Response. The forum allowed for an intersectoral discussion on DRFI among ministries of finance, insurance supervisors, and national DRM agencies from all 10 member states. Key findings and options for recommendations were published in a report, "ASEAN: Advancing Disaster Risk Financing and Insurance in ASEAN Member States: Framework and Options for Implementation" (April 2012), with the endorsement of the Secretary General of ASEAN and the World Bank East Asia and the Pacific Region and Sustainable Development Network Vice Presidents.

countries—the Marshall Islands, Samoa, the Solomon Islands, Tonga, and Vanuatu—are participating in a catastrophe risk insurance pilot for 2012 and 2013 to test the viability of sovereign parametric catastrophe insurance in the region.

Financing challenges for disasters are typically even more acute at the subnational level, although efforts are under way in some countries to improve effectiveness of local financing mechanisms. Local governments typically rely on resource transfers from the central government, both for reconstruction and for other purposes. In some countries, however, local governments are taking action to increase their financial resilience to disasters. In Indonesia, some subnational governments have been insuring public assets since the early 2000s; the municipality of Yogyakarta started insuring its assets in 2003 and received a US$0.5 million payout following the 2006 earthquake to help restore its schools, hospitals, traditional market places, and motor vehicles (World Bank 2011). In 2010 the Philippines passed a new DRM law that revamped its National and Local Calamity Funds into Disaster Risk Management and Recovery Funds and overhauled the budgetary arrangements and requirements of the funds. The Asian Development Bank is also working with Indonesia, the Philippines, and Vietnam to launch disaster finance programs in two megacities in each country, based on urban risk profiling.

Contingency budgets and/or reserve funds often provide an important source of short-term disaster financing at the municipal and provincial levels. Local governments are required or expected to include in their annual budget a contingency budget or reserves for unforeseen events. These funds are the first funds to be accessed in the case of a disaster, with provincial and then central contingency budgets accessed only in the case that the local contingency budget is exhausted or post-disaster spending needs surpass a certain proportion of the local contingency budget. In Vietnam, for example, central and local governments are required to allocate between 2 and 5 percent of their total planned budget for contingent spending on disasters and issues of national defense and security. Once the contingency budgets are exhausted, the government can access Financial Reserve Funds held at the provincial and central levels. In Indonesia the municipal government is charged with developing the post-disaster recovery plan and financing it out of its own budget (through a contingency budget and/or post-disaster budget reallocation); in the case that this budget exceeds 20 percent of their total budget, the municipal government can request support from the provincial or central government. Although contingency budgets and reserves can be an important first source of funds for local governments, experience also shows that they can be difficult to increase or even maintain given the limited size of local governments' revenues and competing demands for their resources.

Legal and administrative challenges remain for the implementation of effective disaster risk financing. In some countries national audit laws prohibit the payment of premiums for risk transfer. In Indonesia, for example, efforts are under way to reform a 2004 law that mandates that a good or service must be received before payment can be made, which prohibits insurance purchase, where a premium is paid before a benefit is realized. In the Philippines, a similar issue had to be resolved to allow the government to procure a Cat-DDO. Beyond securing ex ante financial instruments, concerns remain about countries' ability to effectively appropriate and execute funds following an event. Budget decisions to make resources available are quickly rendered fruitless by the multiple steps required to appropriate and execute those funds. In this context, the administrative and legal dimension of disaster risk financing is as important as the strategy itself. In the Pacific catastrophe insurance pilot, for example, participating countries must address this issue by determining with the World Bank a preagreed mechanism to manage and execute any insurance payouts resulting from the pilot.

Catastrophe Risk Insurance Market: What Is the Insurance Sector's Share in Disaster Losses?

Catastrophe losses in East Asia and the Pacific are primarily borne by the population and the government. In 2011, excluding the floods in Thailand, only 2.1 percent of reported catastrophe losses in East Asia and the Pacific were insured. The region's catastrophe risk insurance markets are characterized by low penetration; non-life insurance penetration in almost all countries for which statistics are available falls below the regional average for Asia (1.59 percent) (figure 6.7). This average, in turn, is well below the European and North American averages of 3.01 and 4.41 percent, respectively. In general, wealth and insurance correlate across East Asia and the Pacific such that the lower the gross national income per capita, the lower the non-life insurance penetration (figure 6.8). The rate of growth of many non-life markets in the region, however, remains strong. This is particularly true in many ASEAN member states, where the market grew 7.1 percent in 2011.[4] One notable exception is China, where penetration declined, likely because of slowing economic growth.

The 2011 floods in Thailand demonstrated the potential for extreme catastrophe losses from flooding in the region as well as the importance of a robust insurance market to absorb catastrophe losses. Damages from the floods, US$46.5 billion, totaled 12.7 percent of Thailand's GDP.[5] Of these, 40 percent (US$12 billion) were insured, with US$3.5 billion of losses falling to the Thai market. Insured losses were primarily driven by the inclusion of flood risk in

Figure 6.7 Non-Life Insurance Penetration in Selected Countries in East Asia and the Pacific and Regionally, 2011

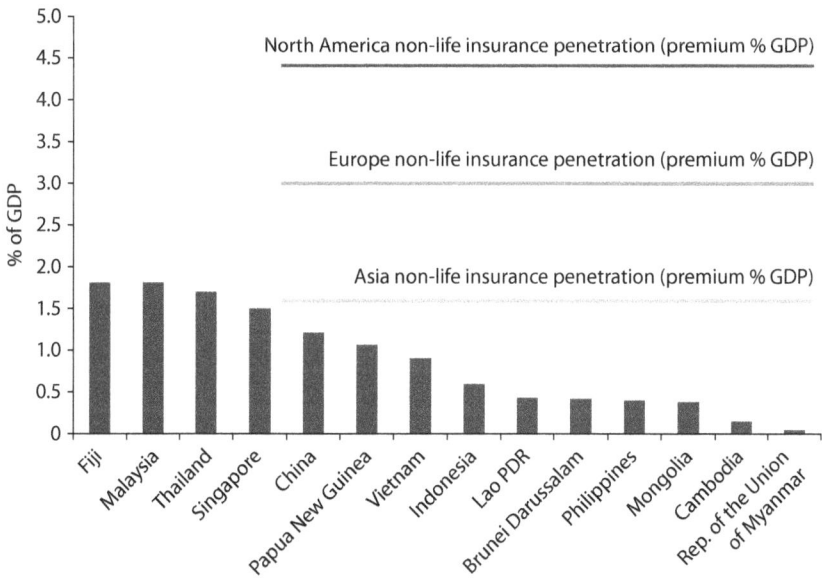

Source: World Bank Disaster Risk Financing and Insurance Program, with data from Swiss Re and AXCO.

Figure 6.8 Non-Life Insurance Penetration versus Gross National Income (GNI) per Capita, 2011

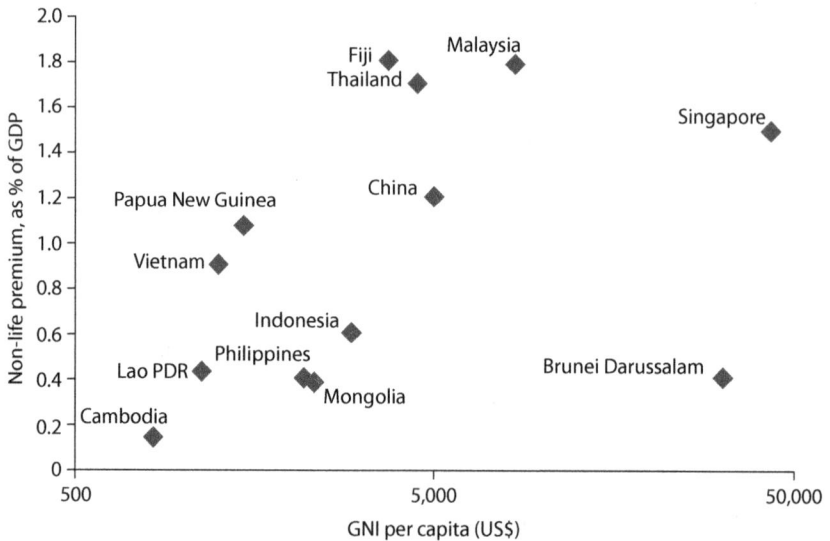

Source: World Bank Disaster Risk Financing and Insurance Program, with data from World Bank, Swiss Re, and AXCO.

industrial all-risk insurance policies, resulting in an incredible loss ratio to insurers of over 3,200 percent for this business. For residential homes and small businesses, flood insurance penetration in Thailand is very low, about 1 percent. It is primarily these losses and public sector losses that comprise the remaining US$18 billion of losses (Swiss Re 2012a, 2012b). The floods carry many important lessons for other countries in East Asia and the Pacific, of which two are highlighted here: first, increasing risk awareness, risk reduction, and insurance uptake by the population is essential in protecting residents against losses; and, second, detailed disaster risk information and insurance premiums based thereon are essential for a performing insurance industry. Insurance supervisors in East Asia and the Pacific have an essential role to play in ensuring the realization of this second takeaway.

Although agriculture is an important socioeconomic sector across countries in East Asia and the Pacific, agricultural insurance has achieved little penetration in the region to date (excepting China). In countries in East Asia and the Pacific where agricultural insurance is available, penetration tends to be about or below 0.01 percent (ratio of premium to agricultural GDP), which is indicative of the limited size of agricultural insurance markets in countries in the region (figure 6.9). In countries where agricultural insurance is available, governments tend to provide significant support to the sector; recently there has been a trend across countries to move toward public-private partnership (PPP) models to improve the sector's performance. In ASEAN countries, agricultural insurance is present in five countries: Indonesia, Malaysia, the Philippines, Thailand, and Vietnam. In the Pacific, there is little evidence of agricultural insurance market

Figure 6.9 Agricultural Insurance Premium, 2009
% of total

Other Asia-Pacific (7)
1%

Australia
4%

ASEAN countries (5)
0%

Korea, Rep.
3%

India
11%

China
50%

Japan
31%

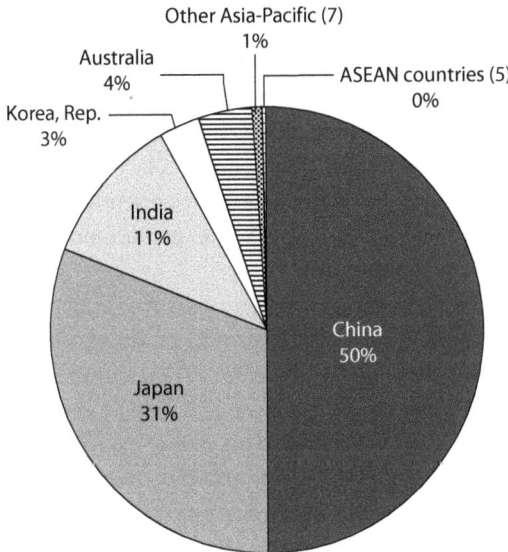

Source: World Bank 2012a.

development in countries for which this information is available (Mahul and Stutley 2010). In Mongolia, the Index-Based Livestock Insurance Program (IBLIP) is spurring the development of the livestock insurance market. The IBLIP is a PPP to increase access to livestock mortality insurance to protect herders against harsh winter weather conditions; launched in 2005, the program is expanding to cover all 21 provinces of Mongolia in 2012 (see box 6.4). The Chinese market, fueled by high-premium subsidies, has expanded rapidly since the mid-2000s and is now the second largest in the world. Although penetration of agricultural insurance (0.40 percent) is still well below the average for advanced economies (1.99 percent), it is above average for middle-income countries (0.29 percent) (Mahul and Stutley 2010).

Disaster microinsurance[6] for low-income populations is as-of-yet undeveloped in nearly all countries in East Asia and the Pacific. International experience suggests that credit-life and life products appear in early stages of microinsurance development, and in time, product diversification occurs and disaster microinsurance is more likely to become available. Although, in some countries in the region, microinsurance markets, particularly for life microinsurance, exist and are growing, the only available evidence of the availability of disaster microinsurance is in China, Indonesia, and the Philippines, where very limited numbers of people are accessing coverage (World Bank 2012a).

Box 6.4 Mongolia Index-Based Livestock Insurance Program

The Mongolia Index-Based Livestock Insurance Program (IBLIP) was launched in 2005 as a PPP with domestic insurance companies to offer affordable and cost-effective livestock mortality insurance to herders, while protecting domestic insurers against major losses that could jeopardize their business. The insurance product is based on an index of livestock mortality rates by species and sum (by county) compiled and maintained by the Mongolian National Statistics Office. Herders bear the cost of small losses (less than 6 percent livestock mortality rate). Larger losses are transferred to the private insurance and reinsurance industry. The final layer of catastrophic loss is borne by the government.

The IBLIP has been successfully scaling up since its launch in 2005. In 2010–11, about 7,000 herders purchased coverage in nine of Mongolia's 21 provinces, expanding to almost 11,000 herders in 15 provinces for the 2011–12 season. For 2012–13, coverage is available nationwide. The largest payouts made by the program were following a devastating 2009–10 season, when 22 percent of Mongolia's livestock died because of the harsh winter; payments to participating herders totaled nearly US$1.3 million. IBLIP's partners are continually improving the program to enhance its performance and facilitate its scaling up, for example, by improving the program's financial resilience by securing reinsurance for catastrophic losses from the international reinsurance market.

Where Do We Want to Be?

There is a need for governments in East Asia and the Pacific to better understand, manage, and reduce the financial and fiscal impacts of disasters. Although there is no "one size fits all" approach in such a diverse region, actions can be taken by all countries in East Asia and the Pacific to improve their financial resilience to disasters. Governments should take a dual, complementary approach. At the sovereign level, they can implement financial strategies to increase their financial response capacity in the aftermath of a disaster while protecting their long-term fiscal balance. Governments can also support the development of catastrophe risk insurance markets to transfer financial risks to the insurance sector to promote the financial resilience of its citizens and the private sector.

Closing gaps in financial and fiscal risk assessment is the starting point for engagement in DRFI in many countries in East Asia and the Pacific. Robust financial and fiscal risk assessment allows for the development of appropriate, cost-effective DRFI strategies. For the government, quantification of its contingent liability to disasters allows it to understand the potential budgetary impacts of these events and to take a proactive approach to reducing budget volatility from disasters. Currently, comprehensive financial risk profiles are available for PICs and under way in Indonesia and the Philippines. In ASEAN, preliminary financial and fiscal risk profiling has been completed, but additional work is required. In China and Mongolia this work has yet to be undertaken.

Integrating disaster risk into fiscal risk and public debt management would allow governments in East Asia and the Pacific to take a holistic approach to managing risk. Fiscal disaster risk assessment reveals the recurrent costs of disasters to the government as well as potential extreme losses due to major events. With this information, the government can develop cost-effective national disaster risk-financing and insurance strategies, including appropriate annual budget allocations for potential disaster events and disaster risk transfer components (such as insurance). These strategies should aim to (1) manage the budget volatility potentially associated with disasters and (2) provide insurance coverage against natural disasters for key public assets. Although efforts are under way in Indonesia, the Philippines, and some PICs to draft national DRFI strategies, currently, no country in East Asia and the Pacific has such a strategy in place. National DRFI strategies would be critical components to informing decision making on DRFI in the region.

Developing catastrophe risk insurance market infrastructure as public goods would increase the insurance sector's share in disaster losses in East Asia and the Pacific. Catastrophe insurance markets in countries in East Asia and the Pacific are generally hampered by low risk awareness of consumers, limited technical capacity and risk-bearing ability of insurers, and/or weak supervisory capacity. The development of catastrophe risk insurance market infrastructure, particularly in middle-income countries in the region, would catalyze market growth by building domestic insurers' capacity while supporting the sale of reliable, cost-efficient insurance products. Catastrophe risk market infrastructure such as product development platforms, catastrophe risk assessment and pricing methodology, and underwriting and loss adjustment procedures provide for standardized product design, reducing the complexity and cost of selling catastrophe insurance products. These tools also help insurers to better understand and monitor accumulations of catastrophe risk in their portfolios. The government can also improve distribution channels, helping insurers reach consumers at the lowest possible cost.

Regional disaster risk-financing and insurance initiatives should promote the standardization of methodologies and products. Building on the experiences with regional cooperation on DRFI in ASEAN countries and PICs, further regional DRFI initiatives would provide multiple benefits. First, such initiatives would provide capacity building and allow for experience sharing across countries. Aligned with this approach, regional initiatives could promote the advancement of standardized methodologies for financial and fiscal risk assessment, facilitating improved understanding of transboundary risks and regional cooperation on developing disaster risk-financing strategies. Regional approaches could also contribute to standardization of products for sovereign disaster risk financing and facilitate risk pooling between and across countries. Finally, regional initiatives could foster regional catastrophe insurance market development and integration. A regional platform, for example, on DRFI in East Asia and the Pacific, could advance these efforts within the region and provide a model for regions around the world.

What Needs to Be Done?

1. **Financial disaster risk assessment and modeling.** There is a need to further develop catastrophe risk modeling tools to assist the countries in East Asia and the Pacific, and particularly ministries of finance, in assessing the economic and budgetary impact of disasters. The development of a standardized exposure database of (public and private) assets at risk is of particular importance.

2. **Sovereign disaster risk-financing strategies for governments.** Such strategies rely on a combination of domestic reserves, contingent credit, and market-based risk transfer mechanisms to ensure immediate access to budget support in the aftermath of a disaster. They also include an insurance strategy of critical public buildings and infrastructure.

3. **Supporting the development of private catastrophe risk insurance markets.** Market-based property catastrophe risk insurance solutions, particularly in middle-income countries, would allow governments to reduce their contingent liabilities associated with disasters by transferring catastrophe risks of residential assets to the private insurance industry. Governments can support the development of risk market infrastructure to allow domestic insurers to underwrite these complex catastrophe risk insurance products in a competitive and sustainable way. This includes the development of an exposure database, product design, and regulatory framework. This can also be facilitated by the establishment of catastrophe risk insurance pools to access international reinsurance markets.

4. **Further promotion of regional cooperation on disaster risk financing and insurance.** Although some countries in East Asia and the Pacific have already made significant progress on DRFI, including Indonesia and the Philippines, there is still a need to promote DRFI and make it an integral part of the DRM agenda. A regional approach should be further explored for risk information systems, standardized products, and capacity building. Such a regional approach is being successfully developed in the Pacific and could be used as an example for other countries in East Asia and the Pacific.

How Can the World Bank Help?

Disaster risk financing and insurance is a long-term effort requiring commitment from countries in East Asia and the Pacific, the World Bank, and other partners at the international and country levels. DRFI is highly technical, requiring the development of specialized technical expertise within the Ministry of Finance, and necessitates cooperation between the Ministry of Finance, national DRM agency, and line ministries. With commitment to develop these areas by national governments, the World Bank has a strong role to play as technical advisor as well as financier for DRFI in the region. The World Bank has an established track record in the field of DRFI and has developed an in-house team of DRFI experts. Furthermore, regional entities, such as the Asian Development Bank and others, have complementary roles to play.

The World Bank can help countries in East Asia and the Pacific develop regional and/or subregional strategies for DRFI that take into account the differential needs of countries. Building on the recent experience with the ASEAN countries, the regional and/or subregional strategies for DRFI engagement would be developed by the DRFI Team, the East Asia and Pacific DRM Team, and the World Bank Treasury, with inputs from Social Protection, Poverty Reduction and Economic Management, and Agricultural Risk Management colleagues. The strategy would define objectives and propose an approach for World Bank engagement in DRFI for the medium term in East Asia and the Pacific. Key action items in the regional strategy could focus on the following:

Development of standardized methodologies, including risk information systems and catastrophe risk modeling, for financial and fiscal disaster risk assessment at the national and subnational levels. The World Bank and GFDRR could invest in the development of national exposure databases (with detailed information on public and private buildings, infrastructure, crops, etc.) and consequence databases (with fiscal and financial data on past disaster events) that are critical components of financial disaster risk assessment. An important role for the World Bank in this effort would be to promote a standardized approach to support the utility and availability of cross-border disaster risk data. Based on these inputs, the World Bank and GFDRR could invest in the development of national catastrophe risk models that would provide the governments with tools to quantify the financial and fiscal impacts of disasters. These efforts would inform key decision makers on the adverse impacts of disasters on the government's budgetary and fiscal position.

Provision of policy advisory services for the development of integrated DRFI strategies, as part of fiscal risk and public debt management. The World Bank could start a dialogue with Ministries of Finance on financial protection of the state against natural disasters through reviews of fiscal management of natural disasters. Detailed reviews of governments' budget management of disasters would help to identify potential short-term resource gaps to finance post-disaster emergency and early recovery as well as longer-term resource gaps to finance post-disaster reconstruction. Based on these reviews, the World Bank could advise Ministries of Finance on the development of National Disaster Risk Financing Strategies. Such strategies would rely on a combination of ex ante financing instruments (e.g., reserves, contingent credit, risk transfer) and ex post financing instruments (e.g., budget reallocation, debt) to secure adequate post-disaster liquidity at the lowest cost. Advisory services should also focus on ex post execution to ensure timely mobilization and strong governance of funds.

Promotion of the development of catastrophe risk market infrastructure, including catastrophe risk modeling, product development platforms, and other information technology tools, to increase catastrophe and weather risk transfer to the private sector. The World Bank and GFDRR could invest in catastrophe risk market infrastructure as public goods, which would simplify the provision of these complex products, and enable insurers to provide cost efficient, affordable, and effective products. In addition, the World Bank could provide advisory services to insurance regulators to move toward regulation that would support the growth of sustainable catastrophe risk insurance markets. Advisory services should focus on developing regulatory regimes that would control exposure to catastrophe risk through a risk-based approach.

Promotion of regional and national DRFI initiatives, including regional risk transfer mechanisms, for increased access to international reinsurance and capital markets and realization of cost savings through economies-of-scale in operating costs and risk pooling. The World Bank could provide advisory services and act as an interface between countries in East Asia and the Pacific, subnational governments, and the international reinsurance and capital markets. In collaboration with countries, the World Bank could develop standardized and replicable trigger mechanisms (building on the outputs of financial risk assessments) as well as design prototype disaster risk financing products. The World Bank could also support the exploration of subregional risk insurance facilities that would increase countries' access to international reinsurance and capital markets and provide cost savings on risk transfer. In addition, the World Bank could provide advisory services on insurance of public assets, which could include review of current policies, improvement of information on assets and of policy terms and conditions, and analysis of group policy placement.

Notes

1. The authors acknowledge that the case of China is unique in the region, because its economy, population, and land mass make it very different from other countries in the region; the World Bank is not currently engaged in disaster risk-financing related topics there, and the authors' understanding of the status of DRFI in this country is, without further analysis, limited.

2. ASEAN member states include Brunei Darussalam, Cambodia, Indonesia, Lao PDR, Malaysia, the Republic of the Union of Myanmar, the Philippines, Singapore, Thailand, and Vietnam.

3. It is noted that the methodology for financial risk profiles of PICs and ASEAN member states is different; the risk assessment conducted for PICs is much more detailed than the preliminary financial risk profiles completed for ASEAN member states. It is also noted, however, that triangulation of results from the regional assessment for ASEAN member states and preliminary financial disaster risk profiles for Indonesia and Vietnam suggest consistency of results across these assessments.

4. Non-life premium growth from Swiss Re Sigma, 2011 data (inflation adjusted) for Indonesia, Malaysia, the Philippines, Singapore, Thailand, and Vietnam. Data for remaining countries are unavailable.

5. Note that figures differ. This report uses the figures from World Bank (2012b) and EM-DAT/UNISDR Annual Statistics 2011 data on percentage of GDP.

6. For the purpose of this report, disaster microinsurance refers to a non-life cover for property, financial assets, or livelihoods that is specifically designed to pay out upon occurrence of a natural disaster.

References

Cummins, D., and O. Mahul. 2010. *Catastrophe Risk Financing in Developing Countries: Principles for Public Intervention*. Washington, DC: World Bank.

EM-DAT. 2012. The International Disaster Database. Centre for Research on the Epidemiology of Disasters (CRED). Accessed September 2012. http://www.emdat.be/database.

Mahul, O., and C. Stutley. 2010. *Government Support to Agricultural Insurance: Challenges and Options for Developing Countries*. Washington, DC: World Bank.

Pacific Catastrophe Risk Assessment and Financing Initiative. 2011. *Risk Assessment Methodology*. Washington, DC: World Bank.

Swiss Re (Swiss Reinsurance Co.). 2012a. *Natural Catastrophes and Man-Made Disasters in 2011: Historic Losses Surface from Record Earthquakes and Floods*. Sigma, Zurich.

————. 2012b. *World Insurance in 2011: Non-life Ready for Take-off*. Sigma, Zurich.

World Bank. 2010. *Weathering the Storm: Options for Disaster Risk Financing in Vietnam*. Washington, DC: World Bank.

————. 2011. *Indonesia: Advancing a National Disaster Risk Financing Strategy—Options for Consideration*. Washington, DC: World Bank.

————. 2012a. *ASEAN: Advancing Disaster Risk Financing and Insurance in ASEAN Countries: Framework and Options for Implementation*. Vols. 1 and 2. Washington, DC: World Bank. http://www.gfdrr.org/gfdrr/sites/gfdrr.org/files/documents/DRFI_ASEAN_REPORT_June12.pdf.

————. 2012b. *Thai Flood 2011: Rapid Assessment Report for Resilient Recovery and Reconstruction*. Washington, DC: World Bank.

World Bank Disaster Risk Financing and Insurance (DRFI) Program. 2012. http://www.gfdrr.org/gfdrr/DRFI.

CHAPTER 7

Sustainable Recovery and Reconstruction

Key Messages for Policy Makers

- The strength of post-disaster recovery and reconstruction efforts lies with how well they respond to the socioeconomic needs of affected people. A deeper understanding of the social impacts of disasters can lead to more responsive and cost-effective rehabilitation programs and a faster recovery and overall reconstruction.
- The quality and effectiveness of disaster response benefits enormously from earlier disaster risk management (DRM) arrangements.
- The reconstruction process offers an opportunity to mainstream exposure, vulnerability, and residual risk reduction into programming as well as to build resilience. Resilience can be built into all reconstruction activities with the overall aim to create communities that are able to withstand and recover quickly from the impact of disasters.
- Key challenges in recovery in East Asia and the Pacific include funding gaps, weak institutional capacity, and making reconstruction and development inclusive of the needs of the most vulnerable populations.

Where Are We Now?

In East Asia and the Pacific, key challenges in achieving an effective and inclusive disaster reconstruction and recovery include the following: funding gaps, inadequate policy framework and institutional capacity, and failure to make the needs for effective reconstruction and development framework inclusive of vulnerable populations. Although relevant to all countries, funding gaps and limited capacity to identify immediate needs are particularly challenging for small and low-income countries as well as Pacific island countries that are fiscally highly

This chapter was written by Zuzana Stanton-Geddes and Patricia Fernandes, with input from Paul Procee and Shyam KC.

exposed to disasters and with a high proportion of vulnerable populations exposed to natural hazards.

Financing recovery and reconstruction remains a heavy burden for economies in East Asia and the Pacific. Successful recovery depends on speedy mobilization and effective disbursement of funds. Inadequate disaster-financing arrangements have exacerbated the adverse socioeconomic consequences of disasters.[1] Disasters place a significant fiscal burden on many governments in the region, and many of them face difficulties in securing adequate and timely funding for early recovery. The cost of recovery and reconstruction varies widely: The 2009 West Sumatra earthquake in Indonesia was estimated at US$2.4 billion, of which almost a third fell on the public sector; Typhoon Ketsana and a second typhoon directly following resulted in recovery and reconstruction requirements totaling US$4.4 billion in the Philippines alone, including US$2.4 billion in public spending needs (World Bank 2012). Cambodia, the Lao People's Democratic Republic, the Republic of the Union of Myanmar, the Philippines, and Vietnam face particularly high annual average expected losses from disasters relative to the size of their economies (World Bank 2012). Cambodia, Lao PDR, and the Philippines could experience bills totaling 18 percent or more of total public expenditure in the event of a 200-year disaster (World Bank 2012). Cambodia, Lao PDR, and the Republic of the Union of Myanmar have in place only very limited budgetary provisions despite their high vulnerability to disasters; these countries regularly struggle to secure adequate and timely funding for early recovery. Fiscal preparation for disasters ranges widely in the region. For example, in Lao PDR, the government does not have a national disaster relief reserve fund to provide funding for emergency response or recovery activities. Funds are mobilized from the national and local budgets in the event of disasters, and the government earmarks a limited amount of the budget for emergency response each year. In contrast, in Indonesia a new government regulation on Funding and Management of Disaster Assistance from 2010 stipulates three categories of funding: a contingency fund, an on-call budget, and social assistance funds that cover a wider range of activities preemptively.

Policy frameworks and institutional capacity constraints persist and hamper effective recovery. Effective reconstruction is set in motion only after the policy maker has evaluated his or her alternatives, conferred with stakeholders, and established the framework and the rules for reconstruction. In East Asia and the Pacific institutions at the national and local levels face capacity, policy framework, human resources, and funding constraints. Disaster recovery depends on coordination among all levels of government, including areas of risk information, standards for informing and guiding disaster recovery strategies, and planning. Post-conflict implications add to fragile institutional frameworks and weak governance and coordination among levels of government and public and private sectors. With the exception of efforts in Indonesia and the Philippines, countries in East Asia and the Pacific do not have well-established systems for routine tracking of public spending on emergency relief, early recovery, and reconstruction. In most countries, the capacity to assess damages and losses is lacking, or it is weak. A strong monitoring system with feedback mechanisms allows institutions and partners to react in real time to the fast-evolving situation on the ground. Prior contracting arrangements, especially in the vital infrastructure sectors such as transport, can help to minimize the gap between relief and recovery efforts. Most countries do not have such arrangements established in their recovery frameworks.

The needs of the most vulnerable populations require greater priority. The impacts on the livelihoods and assets of the poor often go unacknowledged. Living in hazardous environments, the poor are particularly exposed to natural hazards. Lacking assets, savings, or other coping mechanisms, they are also less able to recover quickly. Moreover, disasters tend to have a disproportionately higher adverse impact on women than men because of sociocultural norms and physiological differences. Mortality rates are higher for women than men for various reasons. For example, women represented an estimated 61 percent of fatalities in the Republic of the Union of Myanmar after Cyclone Nargis in 2008, and 70 percent after the 2004 Indian Ocean tsunami in Banda Aceh. Without quality gender-differentiated information to guide planning, existing social and economic inequalities between men and women may not only be reinforced, they may be exacerbated in the reconstruction phase, hampering recovery.

Countries in East Asia and the Pacific have taken considerable efforts to include the impacted communities in the reconstruction and recovery process. Following the 2004 Indian Ocean earthquake and tsunami, in Aceh and Nias, Indonesia, a community-based disaster risk management approach to reconstruction has proven successful to rebuild infrastructure, homes, and the social fabric of affected communities. A community-based resettlement and reconstruction approach was also adopted for housing reconstruction in Yogyakarta and Central Java after the earthquake in 2006. Involvement of women increased accountability and enhanced the appropriateness of technical solutions. This approach resulted in 270,000 earthquake-resistant houses rebuilt in Java within 28 months. A social impact assessment (SIA) provides valuable information and further guidance for programs targeting vulnerable communities, and numerous countries made a social impact analysis part of their post-disaster assessment. The overall challenge is to integrate social protection and livelihood programs in the immediate response and medium- and long-term recovery programs, linking to actions aimed at decreasing vulnerability and exposure of the most vulnerable populations.

Where Do We Want to Be?

Governments can prevent delays in reconstruction and recovery. The recovery process can be severely hampered by the lack of an effective policy environment and leadership, constraints in institutional capacities, insufficient resource mobilization and execution, and a lack of focused efforts to address the immediate and long-term needs of the most vulnerable, who traditionally bear the brunt of disaster impacts. These are areas where governments can make a difference for the lives of people affected when people-focused policies exist and decisions are quickly taken. Evidence highlights the importance of rapidly channeling resources directly to households and communities to minimize delays in the reestablishment of livelihoods. Indebtedness, depletion of assets, and use of negative coping strategies are common post-disaster outcomes, particularly among highly vulnerable populations and households. A key challenge is for governments to take advantage of existing programs with strong community/household outreach, as in the case of community-driven development (CDD) and social protection programs, to better respond to disasters and build resilience from the bottom up. Linking DRM with CDD and social protection programming requires a systematic proactive approach and capacity building for effective integration on the ground.

Use the opportunity to build better structures and infrastructure after a devastating disaster, making recovery a part of—and a model for—sustainable development. The experiences in Indonesia and other parts of the world have demonstrated that a better future is possible even in areas that were poor and downtrodden before the disaster. The guiding principle for every reconstruction is "Don't build the next disaster!" but rather use the opportunity to link restoration livelihoods and economic activity with resilient planning and sustainable development.

Well-designed disaster recovery programs and reconstruction plans rely on an understanding of local realities and channel resources to support the priorities and needs of the affected communities. The main objective of a post-disaster needs assessment (PDNA) inclusive of social needs is to better inform governments' recovery and reconstruction efforts and make them participatory, transparent, and responsive to local needs. The development of a recovery and reconstruction framework is based on a comprehensive estimation of the overall needs for all post-disaster activities. A damage and loss assessment is critical to effectively allocate resources, and a comprehensive results framework adds significant value to reconstruction efforts. Reconstruction planning considers (1) institutional arrangements, (2) financial strategy, (3) community participation, (4) reconstruction strategy, and (5) risk management.

The success of the early and longer-term recovery efforts depends on the ability of programs to fit with the needs and institutions in place in affected areas. Well-designed programs draw on local capacities, understand local realities, and not only address key needs but also strengthen local institutions and practices in ways that enhance development and social cohesion. Understanding how disasters and post-disaster aid efforts affect local livelihood patterns, social structures, and institutions is vital. The failure to tailor recovery activities to the needs of affected groups by missing particular segments of the population, or inequitably distributing support, can result in mistargeted or inadequate forms of support, waste of financial resources, delays in recovery, and in some instances increased social tensions and conflict. There are opportunities for countries in East Asia and the Pacific to systematically use the social impact analysis in the post-disaster phase to shape the design of reconstruction interventions and move from analysis to better results on the ground for affected communities, as well as to take advantage of existing programs with strong community and household outreach to decrease risks and build resilience.

What Needs to Be Done?

1. **Enhance financial resilience through a mixture of adequate ex ante and ex post mechanisms.** Have in place disaster risk-financing and insurance strategies at the national and subnational levels to manage budget volatility linked to disasters to ensure the rapid disbursement and execution of funds after a disaster. Establish national disaster funds as a financial mechanism to conduct transparent and efficient post-disaster damage assessments of public assets (and low-income housing), mobilize immediate post-disaster funding, and execute the funds in close collaboration with relevant line ministries and public agencies. Disaster funds need to be made available as soon as seasonal forecasts are known. Limiting disbursement to after disasters occur may already be too late, particularly for slow-onset disasters such as droughts. Comprehensive tracking and monitoring systems need to be established for funding and recovery needs.

2. **Use appropriate PDNAs to capture the range of socioeconomic and fiscal impacts of disasters and better shape recovery and DRM programs.** The PDNA has been shown to strengthen coordination, leverage financing for multisectoral recovery, and mainstream DRM. Planners should focus on the most vulnerable groups, assessing their needs and implementing appropriate measures so they are not left out by the recovery and reconstruction process. Engagement of communities and stakeholder participation creates ownership in the reconstruction efforts and ensures long-term success. Community-driven recovery and reconstruction projects are useful mechanisms for achieving sustainable impact on the ground. Use of the cross-sectoral PDNA will inform reconstruction and the recovery process as well as development planning. Whichever methodology is chosen, the PDNA should fit with national needs and should be replicable over time, as has been the case for Indonesia.

3. **Make a social impact analysis part of PDNAs and systematically use it in the design of reconstruction interventions.** A systematic use of the social impact assessment (SIA) in the design of reconstruction interventions facilitates countries in East Asia and the Pacific in moving from analysis to better results on the ground. Through a range of instruments and services, the World Bank can help governments in the region make necessary steps to increase their financial protection, strengthen institutional arrangements and community engagement, and link PDNA with a resilient recovery with the overall long-term objective of a sustainable and resilient development.

4. **Translate analysis into better results in DRM programs.** Standardizing the use of SIA in PDNAs requires a sustained capacity-building effort with a focus on key government teams, local research institutions, and civil society organizations. The Philippines, for example, already has benefited from training; further capacity development is needed for other countries in East Asia and the Pacific to streamline the implementation of post-disaster SIAs.

Although universities and research institutes in several middle-income countries are well equipped to adopt and implement the SIA methodology, additional capacity development efforts may be needed for low-income countries. Regional partnerships and arrangements can help. Evidence from the Republic of the Union of Myanmar, the Philippines, and Thailand shows that the social impact of disasters deserves close attention from governments and partners. With various countries in East Asia and the Pacific experiencing recurrent disasters, understanding coping strategies, access to credit, and levels of debt of affected households and communities, as well as the role of formal and informal institutions in identifying and obtaining support, could provide significant insights for policy makers on how to develop programs that can build resilience and ease data collection requirements in the immediate post-disaster phase.

5. **Promote resilient and sustainable planning and development.** To reap the benefits of effective recovery programs, carry-over arrangements should be made to ensure that (1) the policy mandate, gains, and capacity strengthening resulting from this process translate into mainstreaming of risk reduction in regular development and planning and (2) the recovery experience and best practice is retained within institutions and among the public for future reconstruction and recovery needs. Long-term disaster risk reduction measures should be integrated into all lines of a recovery program.

How Can the World Bank Help?

The World Bank helps countries in East Asia and the Pacific in addressing these challenges through (1) a selection of financial instruments, (2) policy advice, (3) PDNAs including a social impact analysis, and (4) community-based programs.

Emergency Recovery Loans (ERLs) help to finance recovery and reconstruction after a major external shock, such as disasters. They are also used to strengthen the management and implementation of reconstruction and recovery efforts, and to develop disaster-resilient technology and early warning systems to prevent or mitigate the impact of future emergencies. In East Asia and the Pacific, several reconstruction projects are financed through ERLs, such as the China Wenchuan Earthquake Recovery Project (US$710 million; see box 7.1), Samoa Post-Tsunami Reconstruction Project (US$9 million), Tonga Post-Tsunami Reconstruction Project (US$5 million), Typhoon Ketsana Reconstruction in Cambodia (US$20 million), and the Vietnam Natural Disaster Management Project (additional financing).

Contingent financing takes the form of self-standing contingent loans or ex ante emergency components of standard investment operations. Before an emergency, Contingent Emergency Response Components can be embedded in investment operations. They have been made part of investment lending projects in Bangladesh, Grenada, India, Indonesia, Lao PDR, and St. Vincent and the

---

Box 7.1 The Wenchuan Emergency Recovery Loan

The US$710 million loan for the Wenchuan Earthquake Project (WERP) was approved in 2009 and comprises two provincial parts: US$510 million for Sichuan and US$200 million for Gansu, the two provinces hardest hit by the 2009 earthquake. The focus is on the reconstruction of infrastructure such as roads, bridges, water supply, wastewater and solid waste, and health facilities in selected counties in both provinces and, in the case of Gansu, on the reconstruction of education facilities. The project is designed using a flexible framework approach. During project preparation, the Bank agreed on the general framework and principles with the central and the provincial governments, while the specific investments would be identified during project implementation.

Numerous subprojects have been completed in the course of a year and are playing a catalytic role in transforming cities, improving basic services and the quality of life and the environment. The project's impacts are very apparent, and the loan has clearly leveraged additional investments in urban services and infrastructure and private development in cities. The international construction safety standards applied to the Bank-financed infrastructure are now being used for other investments as well.

The government, with support from the World Bank and Global Facility for Disaster Reduction and Recovery (GFDRR), has finalized an overall evaluation of China's reconstruction efforts after the Wenchuan earthquake and is now planning to organize an international workshop to share the China and WERP experience in post-disaster recovery and provide advice on a range of issues, such as effective planning and managing emergency response, improving health services in the aftermath of disaster, and pairing programs to expedite reconstruction and integrated housing and livelihood improvement programs. The Bank is also implementing 17 urban flood risk management projects, focusing mainly on structural measures, with an overall portfolio of nearly US$1.9 billion, out of which US$830 million is Bank-financed.

Source: World Bank staff.

Grenadines. In Lao PDR, following Typhoons Haima and Nok Ten in 2011, the contingent component was utilized to meet immediate emergency road repairs. The Development Policy Loan with Catastrophe Deferred Drawdown Option (Cat-DDO) is a self-standing ex ante contingent financing instrument available for World Bank countries. It offers immediate budget support to cover urgent financing needs in the aftermath of a national disaster while other resources, including national, bilateral aid, or reconstruction loans, are mobilized. The Philippines was the first country in East Asia and the Pacific with a Cat-DDO (amounting to US$500 million), which was fully disbursed following the devastating impacts of Tropical Storm Sendong (Washi).

The Immediate Response Mechanism allows countries to quickly access funds. World Bank International Development Association clients can access up to 5 percent (or up to US$5 million) of the undisbursed balances of the country's portfolio of investment projects. The instrument requires an operations manual and up-front agreements on implementation arrangements and contingency components in place before disbursement. The Crisis Response Window provides limited additional financing, which can be accessed after severe exogenous shocks. Access is linked to country-specific circumstances such as magnitude of the crisis and access to alternative sources of financing.

The World Bank is helping countries in East Asia and the Pacific to improve their fund-tracking mechanisms to better respond to the needs of the affected populations. In Indonesia, Reconstruction Expenditure Tracking Analysis Methodology was developed in the aftermath of the 2004 earthquake and tsunami to track financial progress. A joint team of the Reconstruction Agency and the World Bank tracked flows since the beginning of the recovery effort to establish geographical and sectoral gaps and identify underfunded regions and sectors. Based on these gap assessments, the government of Indonesia and the Multidonor Fund allocated additional funds to close the gaps (World Bank 2011a). In the Philippines, following Tropical Storm Ondoy and Typhoon Pepeng in 2009, the Philippine government started an initiative to track assistance from international partners, local civil society organizations, and the private sector as well as from government. The next step is to institutionalize them into more permanent systems, as already planned in the Philippines.

The World Bank and GFDRR assist disaster-hit countries in assessing medium- and longer-term disaster impacts and recovery needs for sustainable recovery.[2] Conducted immediately after the disaster, the PDNA is a government-led exercise that estimates the economic impact of the disaster and determines needs for resilient recovery. The PDNA provides a baseline for external and domestic resource mobilization for recovery and forms the basis for the development of a well-defined recovery framework that ensures needs are prioritized and sequenced correctly. The PDNA process has proven useful in strengthening coordination, leveraging financing for multisectoral recovery, and mainstreaming DRM into post-disaster sectoral work (box 7.2). For example, as a direct result of the PDNA in Manila following Typhoons Ondoy and Pepeng, the government is working with the World Bank to prepare a Flood Management Master Plan for metropolitan Manila. In Lao PDR, capacity building provided by the World Bank following the devastating Typhoon Ketsana in 2009 made a significant contribution to the government's ability to assess damage and losses in the aftermath of Tropical Storm Haima in 2011, allowing previously trained government officials to lead a speedy, high-quality assessment.

The World Bank is helping countries in East Asia and the Pacific to include a qualitative analysis of the social needs as part of the post-disaster needs assessment. Using mostly qualitative, field-based methods, the SIA can rapidly generate information on critical issues that would otherwise remain hidden (box 7.3). The SIA has been instrumental in highlighting cross-cutting issues, such as social accountability and the use of negative coping strategies that do not fit neatly within one particular sector, the perspectives of affected communities and their key priorities, community dynamics and how these affect recovery, and identification of whether particular social groups or areas are (at risk of) being

Box 7.2 Samoa Tsunami Post-Disaster Needs Assessment and Resilient Reconstruction

On September 29, 2009, a powerful earthquake struck the south of Samoa and was followed by a tsunami that impacted American Samoa, Samoa, and northern Tonga. In Samoa there were 143 reported deaths, and an estimated 5,274 people were made homeless. The main affected areas, with about 7 percent of the country's population, were the southern, eastern, and southwestern coast of Upolu Island. Within two weeks of the disaster the government of Samoa launched a PDNA with the World Bank in the technical lead and a team made up of UN and Asian Development Bank partners.

The assessment estimated the sum of direct damages and economic losses in Samoa at US$124 million, equivalent to 22 percent of Samoa's gross domestic product. The greatest costs were found in the transport sector, where the coastal road and accompanying sea walls were severely damaged. About 13 percent of the country's housing stock was lost. A significant proportion of the affected population had chosen to relocate inland from the shore, and the government of Samoa decided to support this move with the provision of services. This move inland, as part of the recovery and future disaster risk reduction, raises the total post-tsunami recovery costs to about US$167.4 million. Results from the PDNA were used by the Ministry of Finance to prepare the midyear supplementary budget and the post-tsunami recovery program.

Source: World Bank staff based on World Bank, GFDRR, and Government of Samoa 2009.

Box 7.3 What Is a Post-Disaster Social Impact Analysis?

In the post-disaster context, social impact analysis is narrowly defined in the context of disaster recovery.[a] It focuses on the social and socioeconomic aspects of people's lives most closely connected to their efforts to rebuild their lives and communities. This includes how people manage the collective challenge of recovery, how the disaster and aid effort affect the assets and capabilities of different socioeconomic groups and their ability to recover their livelihoods, and how the disaster and aid effort affect social relations and community institutions. The approach uses standard tools and methodologies (participant observation, focus group discussions, and key informant interviews) adapted to the post-disaster context. SIA takes into account the existing constraints in financial resources, time, and local research capacity to ensure that data can be collected within the tight time frame of a standard PDNA. Research can usually be organized in four areas:

Socioeconomic Impacts: How do the disaster and aid effort affect the assets, capabilities, and ability to recover of different social groups? This includes impacts on the local socioeconomic structure, including how people work and earn a living, impacts on people's access to capital, impacts on managing land and other resources, and impacts on how people cope, including through migration.

box continues next page

Box 7.3 What Is a Post-Disaster Social Impact Analysis? *(continued)*

Impacts on Social Relations and Cohesion: How do the disaster and recovery effort affect social relations at a community level? This includes impacts on social capital and cohesion and impacts on the social composition of affected communities and relations among social groups differentiated by factors such as gender, ethnicity, age, religion, and, if relevant, caste.

Relief, Recovery, and Accountability: How do affected communities perceive, participate in, and negotiate their interests regarding the relief and recovery effort? This includes overall patterns of relief and reconstruction support; targeting equity and vulnerability; the process of delivery; decision making; and the resolution of problems related to the implementation of relief and reconstruction efforts.

Community and Institutional Impacts: How do the disaster and recovery effort affect community organizations and the rules, incentives, and social norms that govern how people interact and behave? This includes impacts on relations between community members and leaders and the impact on community and intervillage organizations.

a. For a broad parameter of social analysis, see, for example, IAIA (2003) or World Bank (2003).

---

excluded from rehabilitation and reconstruction efforts. SIAs have been conducted as part of PDNAs in the Republic of the Union of Myanmar, the Philippines (2009), and Thailand (2011). In the Philippines, analysis contained three main focus areas: livelihoods and coping strategies, social relations and cohesion, and local governance and social accountability. The findings, which were centered around governance, social accountability, people's coping strategies, and impacts on vulnerable populations, highlighted key issues that would not have been captured using the standard methodology alone.

The World Bank helps to focus on vulnerable communities and households through community-driven development and innovative social protection interventions. As mentioned in chapter 2, engagement of communities and stakeholder participation creates ownership in the reconstruction efforts and ensures long-term success (World Bank 2011a). In December 2010, the Bank rapidly mobilized US$3.5 million of grants through the Java Reconstruction Fund (JRF) to finance the reconstruction and rehabilitation of housing and community infrastructure that was destroyed by the eruption of Mount Merapi through the existing CDD program (REKOMPAK). JRF/REKOMPAK, the government's flagship poverty alleviation program, focuses heavily on a community-driven approach in all aspects of development, including infrastructure, savings/loans, employment, livelihoods, and post-disaster, post-conflict response (World Bank 2010a, 2010b). Based on the lessons learned from major housing reconstruction programs, *Safer Homes, Stronger Communities: A Handbook for Reconstructing after Natural Disasters* (World Bank 2010c) provides a framework for policy makers and project managers to make the multisectoral decisions involved in major housing and community reconstruction projects that empower communities affected by disasters to build resilience against future vulnerability (box 7.4).

Box 7.4 Safer Homes, Stronger Communities: Guiding Principles

- Reconstruction begins the day of the disaster.
- A good reconstruction policy helps reactivate communities and empowers people to rebuild their housing, their lives, and their livelihoods in a better and safer way.
- Reconstruction has to be sustainable to contribute to long-term development.
- Community members should be partners in policy making and leaders of local implementation.
- Civil society and the private sector play a vital role.
- Reconstruction policy and plans should be financially realistic but ambitious with respect to disaster risk reduction.
- Institutions matter and coordination among them improves outcomes.
- Reconstruction is an opportunity to plan for the future and carry lessons learned from the past.
- Relocation disrupts lives and should be minimized.
- Assessment and monitoring can improve reconstruction outcomes.
- Every reconstruction project is unique.

Source: Adapted from World Bank 2010c.

Linking social protection programs and CDD interventions with DRM can help build climate and disaster resilience hand in hand with the most vulnerable communities. As described in chapter 2, establishing systems that can respond quickly to disasters reduces delays for households in accessing resources to restore livelihoods. Good practice examples from Indonesia and the Philippines, using CDD approaches for post-disaster reconstruction, demonstrate the effectiveness of community-focused approaches when prior investments facilitated the creation of networks and resources to reach out to the populations at the grassroots level. International experience demonstrates that existing social protection programs can be rapidly scaled up (as well as scaled down) to include affected populations in the aftermath of a disaster. Gender disparity is a key concern in post-disaster recovery and reconstruction (box 7.5). An Australian Agency for International Development (AusAID)-funded program on Gender and Disasters, active in Indonesia, the Philippines, the Solomon Islands, and Vietnam, offers analytical and operational guidance to make DRM programs, including CDD programs, post-disaster assessments, and post-disaster recovery, inclusive of gender needs.

The World Bank supports countries to link their recovery efforts with resilient development. The Tonga Post-Tsunami Reconstruction Project supports the construction of cyclone-resistant housing, geographic information system (GIS) and risk assessment capacities, and the preparation of community DRM plans. In Samoa, the Post-Tsunami Reconstruction Project supports disaster-resilient community coastal infrastructure management plans. The World Bank assists the government in its efforts to support the relocation and rehabilitation of communities living on the island of Upolu through a US$10 million International Development Association credit with additional grant co-financing of US$1.79

Box 7.5 Gender Concerns and the Issue of Land Titling

In the post-disaster period, housing reconstruction, land titling, and ownership claims bear important gender dimensions. In Indonesia, in areas where land titling has been carried out, registration data from 1998 shows that 28 percent of titles are in women's names. The Indian Ocean tsunami that struck Indonesia in 2004 affected more than 800 kilometers of coastline and destroyed up to 53,795 land parcels. A World Bank study (2010d), based upon experience from the Reconstruction of Aceh Land Administration System project to support reconstruction of housing and communities in post-tsunami Aceh and North Sumatra, finds that the tsunami disaster put women on the verge of losing livelihoods and assets, because women's land and property rights were not acknowledged uniformly, and that affected women found it difficult to register and secure a title certificate for inherited claims.

Lessons learned: Prior registration of houses and land ownership taking into consideration both the male and female owners is an indispensable step forward in helping with relocation and asset compensation processes. The World Bank has been helping women in many countries to better understand their rights and secure clear land title to their properties. Examples of recent World Bank operations in land titling include Honduras and the Philippines. In the case of a disaster, clear land titling rights help to protect women from losing assets due to lacking documentation or administrative insecurities.

Source: World Bank 2011b.

Box 7.6 Linking Recovery with Resilient Development in Aceh

Through trust-funded activities under the Multidonor Fund for Aceh and Nias and the Java Reconstruction Fund, the World Bank mobilized US$785 million for post-tsunami reconstruction in Aceh and for post-earthquake reconstruction in Yogyakarta. With support from the Global Facility for Disaster Reduction and Recovery, the Bank also managed a total of US$3.25 million for technical assistance to support risk identification, integration of risk reduction in investment projects, and identification of policy options for catastrophic risk financing. To date the above post-disaster reconstruction had built more than 34,000 earthquake-resistant houses, 2,655 kilometers of roads, 936 bridges, and 1,473 irrigation channels. Disaster-contingent components and disaster mitigation have also been included in two World Bank–supported projects (PNPM-Urban III/CDD, US$85 million, and WINRIP/road improvement, US$250 million).

In ex ante risk reduction, Indonesia's central government developed a master plan for reconstruction through the national planning agency. The plan laid out a vision for reconstruction across all sectors, including post-disaster spatial planning, land-use, and related affairs. After the master plan for reconstruction was issued, the central government established the Agency for the Rehabilitation and Reconstruction of Aceh and Nias. This reconstruction agency had a mandate to coordinate and implement reconstruction projects including those by foreign agencies. Working on the principle of new post-disaster spatial plans to

box continues next page

Box 7.6 Linking Recovery with Resilient Development in Aceh *(continued)*

reduce vulnerability of populations in risk-averse areas, a resettlement policy had to be instituted in certain cases. Early in the reconstruction process, the Executing Agency for Rehabilitation and Reconstruction of Aceh-Nias, through the powers of its institutional mandate, provided local authorities with funds to acquire new lands for resettlement, providing the most at-risk families with the means to reduce their vulnerability.

Source: World Bank staff.

---

million from the Pacific Region Infrastructure Facility. The project provides (1) improved road and pedestrian access to relocation areas, (2) restored road access to communities affected through repairs to main and secondary land transport routes and associated seawalls damaged by the tsunami, (3) a more resilient road network that provides continued access to populated areas during and after disasters, and (4) improved planning through updated coastal infrastructure management plans and improved base and hazard mapping. Box 7.6 describes the experience in linking recovery with resilient development in Aceh.

Notes

1. Based on World Bank (2011a): Delayed recovery (1) exacerbates traumatic effects and immediate impacts of a disaster, (2) can cause nonresilient restoration or reconstruction of social, physical, and productive infrastructure and restore the very vulnerabilities that caused the disaster in the first place, and (3) can exacerbate indirect and long-term effects, causing economic slowdown, disruption of regional and global trade patterns, reduced incomes and liquidity in the local economies, unemployment, and other problems, with ramifications ranging from chronic indebtedness to malnutrition to gender inequity. In economic terms, delayed reconstruction adds to the already high cost of impacts on human, economic, and fiscal resources.

2. From 2008 to 2012 the World Bank supported the governments of Cambodia (Typhoon Ketsana), Indonesia (earthquake), Lao PDR (Tropical Storm Haima and Typhoon Ketsana), the Philippines (Tropical Storm Sendong and Typhoons Ondoy and Pepeng), Samoa (tsunami), and Thailand (floods) to conduct PDNAs that formed the basis for reconstruction and recovery and helped mobilize international financial support for reconstruction. After Cyclone Nargis in the Republic of the Union of Myanmar in 2008, the Bank with the UN supported ASEAN in the Post-Nargis Joint Assessment.

References

IAIA (International Association for Impact Assessment). 2003. *International Principles for Social Impact Assessment.* Fargo, ND: IAIA.

World Bank. 2003. *Social Analysis Sourcebook.* Washington, DC: World Bank.

————. 2010a. "First Step toward Post-Merapi Reconstruction." December 29. http://www
.worldbank.org/en/news/2010/12/29/first-step-towards-post-merapi-reconstruction.

————. 2010b. "Measuring Merapi's Losses." November 24. http://www.worldbank.org/
en/news/2010/11/24/measuring-merapis-losses.

————. 2010c. *Safer Homes, Stronger Communities: A Handbook for Reconstructing after
Natural Disasters.* Washington, DC: World Bank. http://www.housingreconstruction
.org.

————. 2010d. *Study on Gender Impacts of Land Titling in Aceh.* World Bank Social
Development Department. Washington, DC: World Bank. http://siteresources
.worldbank.org/INTEAPREGTOPSOCDEV/Resources/Aceh_web.pdf.

————. 2011a. *Haiti Earthquake Reconstruction: Knowledge Notes from DRM Global Expert
Team for the Government of Haiti.* Washington, DC. http://www.gfdrr.org/gfdrr/sites/
gfdrr.org/files/publication/GFDRR_Haiti_Reconstruction_KnowledgeNotes_0.pdf.

————. 2011b. *Integrating Gender Issues in Recovery and Reconstruction Planning.* Gender
and DRM Guidance Notes, Issue 5, Washington, DC.

————. 2012. *Advancing Disaster Risk Financing and Insurance in ASEAN Member States:
Framework and Options for Implementation.* Vols. 1 and 2. Washington, DC: World
Bank.

World Bank, GFDRR (Global Facility for Disaster Reduction and Recovery), and
Government of Samoa. 2009. *Samoa Post-Disaster Needs Assessment Following the
Earthquake and Tsunami of 29th September 2009.* Washington, DC. http://www.gfdrr
.org/gfdrr/sites/gfdrr.org/files/documents/PDNA_Samoa_2009.pdf.

Urbanization by Region

Figure A.1 Growth of Cities

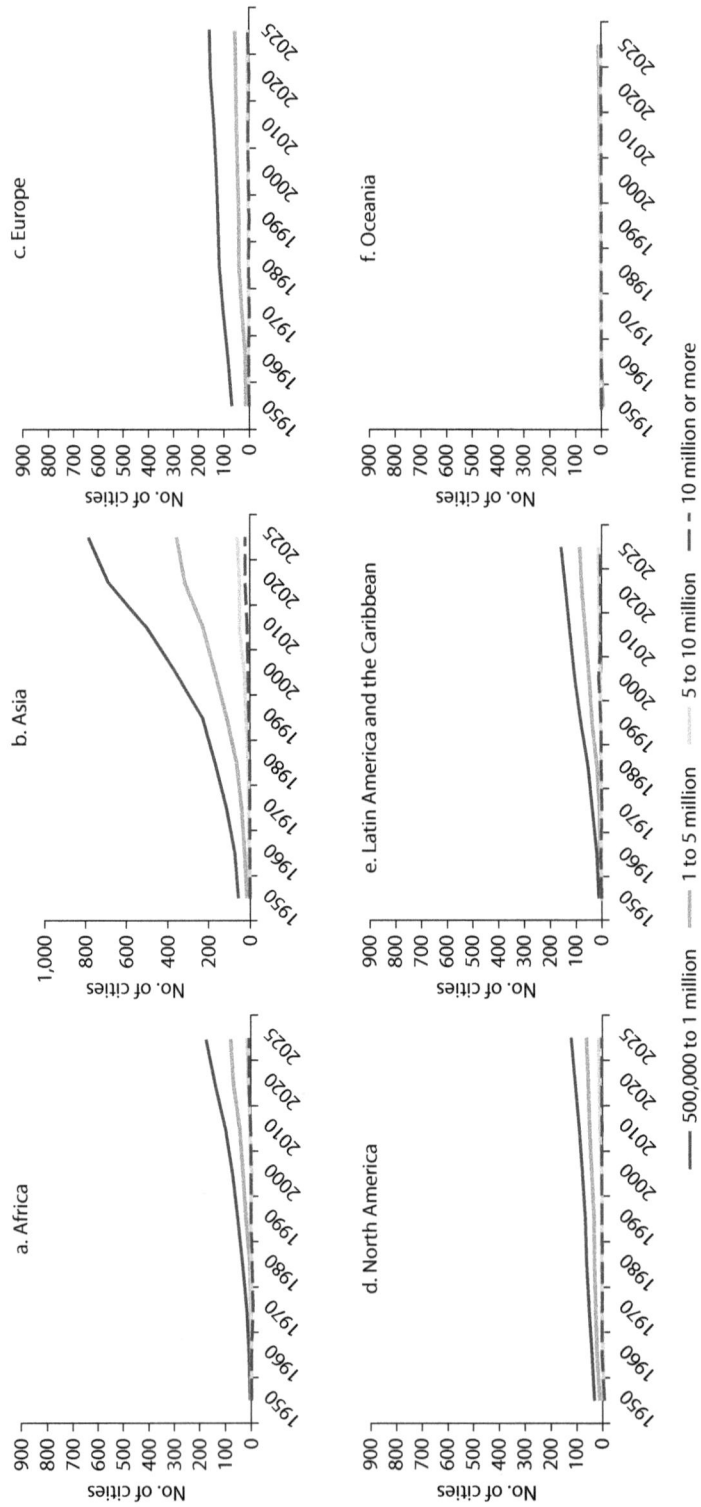

a. Africa

b. Asia

c. Europe

d. North America

e. Latin America and the Caribbean

f. Oceania

No. of cities

—— 500,000 to 1 million   ——— 1 to 5 million   ——— 5 to 10 million   – – 10 million or more

Source: UN DESA (United Nations Department of Economic and Social Affairs/Population Division). 2011. *World Urbanization Prospects: The 2011 Revision.* New York.

Figure A.2 Growth of Urban Population

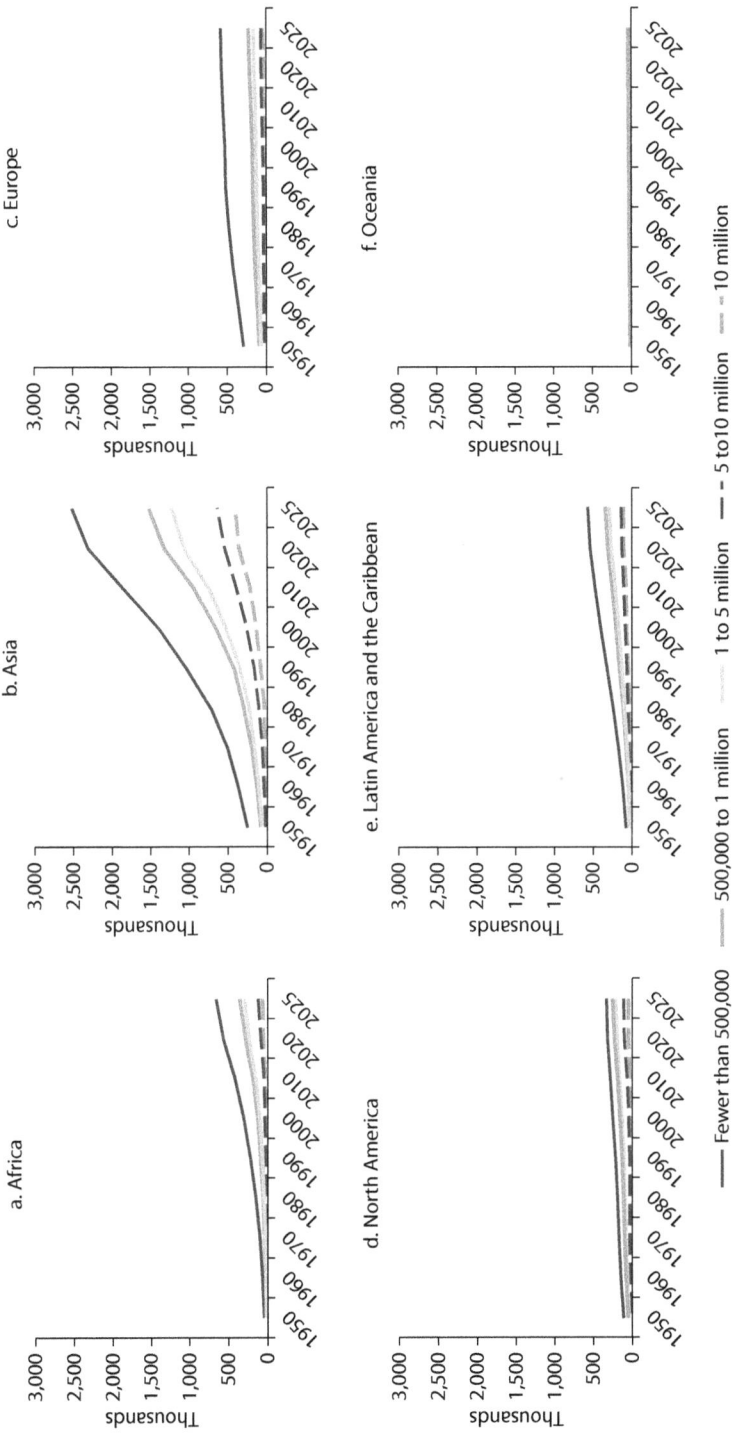

a. Africa

b. Asia

c. Europe

d. North America

e. Latin America and the Caribbean

f. Oceania

Fewer than 500,000 ——— 500,000 to 1 million ——— 1 to 5 million — — 5 to10 million ···· 10 million

Source: UN DESA (United Nations Department of Economic and Social Affairs/Population Division). 2011. *World Urbanization Prospects: The 2011 Revision.* New York.

Large-Scale Disasters in Asia 2008–11

Figure B.1 Large-Scale Disasters in Asia 2008–11

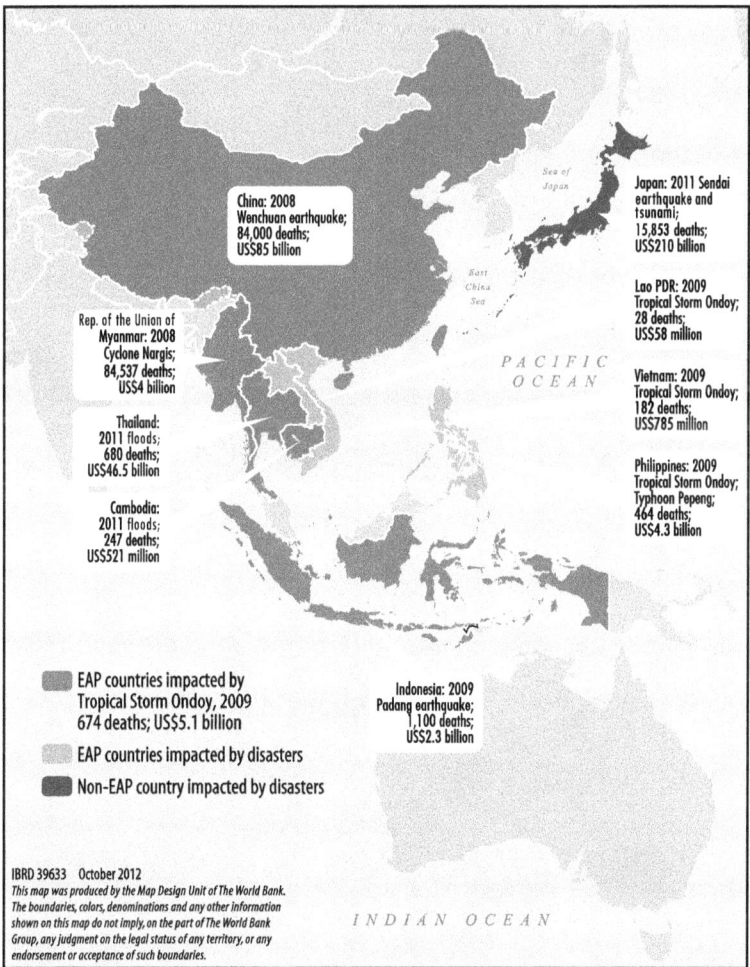

China: 2008
Wenchuan earthquake;
84,000 deaths;
US$85 billion

Japan: 2011 Sendai
earthquake and
tsunami;
15,853 deaths;
US$210 billion

Lao PDR: 2009
Tropical Storm Ondoy;
28 deaths;
US$58 million

Rep. of the Union of
Myanmar: 2008
Cyclone Nargis;
84,537 deaths;
US$4 billion

Vietnam: 2009
Tropical Storm Ondoy;
182 deaths;
US$785 million

Thailand:
2011 floods;
680 deaths;
US$46.5 billion

Philippines: 2009
Tropical Storm Ondoy;
Typhoon Pepeng;
464 deaths;
US$4.3 billion

Cambodia:
2011 floods;
247 deaths;
US$521 million

Sea of
Japan

East
China
Sea

PACIFIC
OCEAN

EAP countries impacted by
Tropical Storm Ondoy, 2009
674 deaths; US$5.1 billion

Indonesia: 2009
Padang earthquake;
1,100 deaths;
US$2.3 billion

EAP countries impacted by disasters

Non-EAP country impacted by disasters

IBRD 39633 October 2012
*This map was produced by the Map Design Unit of The World Bank.
The boundaries, colors, denominations and any other information
shown on this map do not imply, on the part of The World Bank
Group, any judgment on the legal status of any territory, or any
endorsement or acceptance of such boundaries.*

INDIAN OCEAN

Source: World Bank staff based on data accessed from CRED EM-DAT database 2012. The International Disaster Database. Centre for Research on the Epidemiology of Disasters (CRED), Brussels. http://www.emdat.be/database. Accessed September 2012.

Vulnerability of Cities to Multiple Hazards in East Asia and the Pacific

Figure C.1 Cities in East Asia and the Pacific Vulnerable to Multiple Hazards

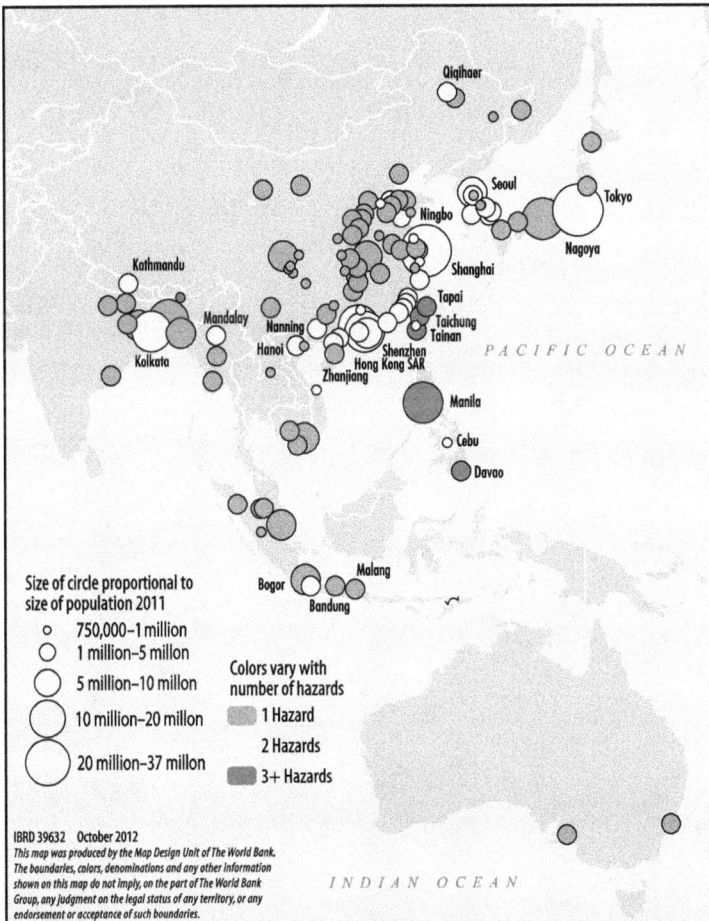

Source: World Bank staff with data from UN DESA 2011 and CIESIN. UN DESA (United Nations Department of Economic and Social Affairs/Population Division). 2011. *World Urbanization Prospects: The 2011 Revision.* New York.

Risk Identification Monitoring

The **Hyogo Framework for Action** (HFA) offers a means of monitoring the status of key disaster risk management indicators in a consistent methodology. Although there are invariably gaps in reported data, the available results from the 2009–11 National Progress Reports (see UNISDR 2011a, 2011b, 2011c) are valuable for assessing the current status in countries in East Asia and the Pacific. The list below highlights the indicators from priorities 2 and 3 that are most relevant for risk identification:

1. Ensure that disaster risk reduction is a national and a local priority with a strong institutional basis for implementation.
2. Identify, assess, and monitor disaster risks and enhance early warning.
   2.1 *National and local risk assessments based on hazard data and vulnerability information are available and include risk assessments for key sectors.*
   2.2 *Systems are in place to monitor, archive, and disseminate data on key hazards and vulnerabilities.*
3. Use knowledge, innovation, and education to build a culture of safety and resilience at all levels.
   3.1 *Relevant information on disasters is available and accessible at all levels, to all stakeholders.*
   3.2 *Research methods and tools for multirisk assessments and cost-benefit analysis are developed and strengthened.*
4. Reduce the underlying risk factors.
5. Strengthen disaster preparedness for effective response at all levels.

The averaged results in figure D.1 are self-reported by the countries on a 1–5 scale with 1 representing "minor" achievement and 5 indicating "comprehensive" achievement. The score averaged across indicators 2.1, 2.2, 3.1, and 3.2 is available for a selection of countries in East Asia and the Pacific to capture a metric

Figure D.1 Indicator Results Related to Risk Identification HFA National Progress Reports

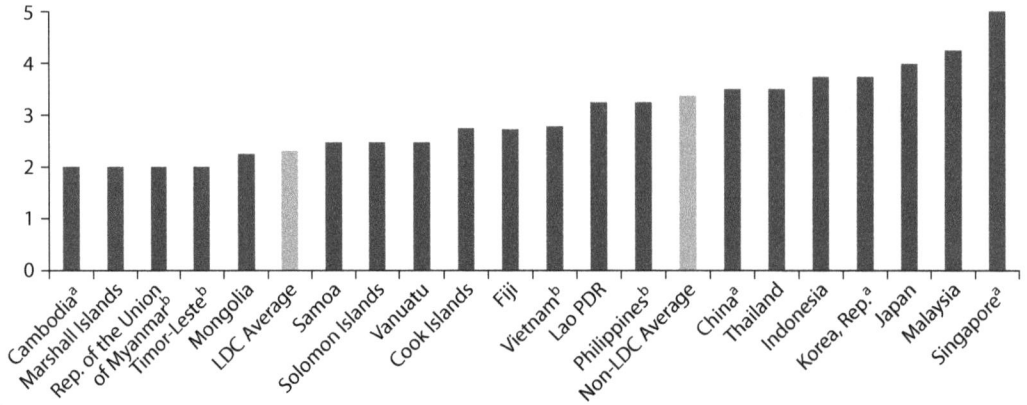

Source: UNISDR 2011a.
Note: LDC = least developed country.
a. Indicates results from 2007–09.
b. Interim reports.

Table D.1 Select Survey Results Related to Risk Identification from HFA National Progress Reports

	Average risk identification indicator (sorted low to high)	2.1 Is there a national multihazard risk assessment available to inform planning and development decisions?	2.2 Are disaster losses systematically reported, monitored, and analyzed?	3.1 Is there a national disaster information system publicly available?	3.3 Is DRR included in the national scientific applied-research agenda/budget?
Marshall Islands	2.00	No	No	No	No
Timor-Leste[a]	2.00	Yes	Yes	No	No
Mongolia	2.25	Yes	No	No	Yes
Samoa	2.50	No	No	No	No
Solomon Islands	2.50	No	No	No	No
Vanuatu	2.50	No	Yes	Yes	No
Cook Islands	2.75	No	No	Yes	Yes
Fiji	2.75	No	No	No	No
Vietnam[a]	2.75	No	Yes	Yes	Yes
Lao PDR	3.25	Yes	No	No	No
Philippines[a]	3.25	Yes	Yes	Yes	Yes
Thailand	3.50	Yes	NR	NR	NR
Indonesia	3.75	Yes	Yes	Yes	Yes
Japan	4.00	Yes	Yes	Yes	Yes
Malaysia	4.25	NR	NR	No	Yes
% yes		46.7	40.0	40.0	46.7

Source: UNISDR 2011a.
Note: DRR = disaster risk reduction. NR = no response.
a. Interim reports.

related to risk identification. General trends show that for the least-developed countries (LDCs), the average result is 2.3 and is a full point lower than the average for non-LDCs. As part of the survey process key questions are asked related to specific indicators. Table D.1 summarizes the results of the questions for the risk identification–related indicators.

References

United Nations International Strategy for Disaster Reduction (UNISDR). 2011a. *The Compilation of National Progress Reports on the Implementation of Hyogo Framework for Action 2009–2011*. New York.

———. 2011b. *HFA Progress in Asia Pacific: Regional Synthesis Report 2009–2011*. New York. http://www.unisdr.org/we/inform/publications/21158.

———. 2011c. *Hyogo Framework for Action 2005–2015: Building the Resilience of Nations and Communities to Disasters, Mid-Term Review 2011*. New York. http://www.preventionweb.net/files/18197_midterm.pdf.

Action Plan for Building Earthquake Resilience

Short Term (as Soon as Possible)

1. Initiate at least one narrowly focused earthquake risk reduction program in a major metropolitan area for maximum impact on potential life and economic losses in the public sector; possibly start with schools and hospitals for life losses, and power generation and distribution systems for economic losses.
2. Assess integration of earthquake risk assessments and risk reduction into all major future infrastructure investments.
3. Review and update existing building codes and their enforcement, specifically for earthquakes.
4. Conduct a critical review of national earthquake risk reduction policies and laws.

Medium Term (the Next 5 Years)

1. Complete one large but narrowly focused earthquake risk reduction program for maximum impact on life losses in the public sector as a demonstration project. The ISMEP program in Turkey is a good example.
2. Complete one narrowly focused earthquake risk reduction program for maximum impact on economic losses in the public sector as a demonstration project.
3. Demonstrate that cost-effective strengthening options are available for vulnerable structures and gain public support; schools are easiest.
4. Redefine the earthquake hazardous areas based on history and additional research, especially on geologic faults.
5. Redefine tsunami hazardous areas; improve tsunami warning systems.
6. Update the codes and add requirements for the strengthening of existing buildings.
7. Strengthen enforcement of the codes and the inspection of construction.
8. Conduct training programs for structural engineers in earthquake risk analysis and risk reduction, including strengthening of existing vulnerable structures. Training programs for contractors and the trades would also be very useful.

9. Mandate professional registration for structural engineers, particularly in the earthquake areas of each country.

Long Term (5–10 Years)

1. Initiate and support long-term earthquake risk reduction programs to impact all key public sectors.
2. Support or initiate long-term earthquake risk reduction programs for the highest risk private structures, which are typically the most vulnerable structures in the region.
3. Support or initiate long-term earthquake risk reduction programs for the highest risk industries and maximum economic impact.
4. Pass legislation to require strengthening of private sector structures and infrastructure with or without public financing but with incentives.

Source: World Bank. 2010. *It Is Not Too Late: Preparing for Asia's Next Big Earthquake, with Emphasis on the Philippines, Indonesia, and China* [Policy Note], by P.I. Yanev. Washington, DC: World Bank.

Classification of Meteorological Services in East Asia and the Pacific

Countries in East Asia and the Pacific are based on World Bank regions. Meteorological services are based on World Meteorological Organization regional membership (including designated NMS [national meteorological service]). The numbers following the World Bank East Asia and the Pacific countries indicate the composite classification of the NMSs according to their capacity to observe and forecast, provide climate services, and deliver services. 0 = no services; 1 = basic services; 2 = essential services; 3 = full services; 4 = advanced services. (See chapter 5 for definitions.)

Countries in the World Bank East Asia and the Pacific region:

Cambodia (1)	Korea, Rep. (4)	Rep. of the Union of Myanmar (1)	Singapore (3)
China (4)	Lao PDR (2)	Pacific island countries (0–1)	Thailand (3)
Indonesia (3)	Malaysia (3)	Papua New Guinea (1)	Timor-Leste (3)
Japan (4)	Mongolia (2)	Philippines (3)	Vietnam (2–3)

WMO Region II (Asia) countries, which overlap with East Asia and the Pacific geographically, as well as the most developed NMSs in WMO region II (denoted by **)

Cambodia (Department of Meteorology)
China (China Meteorological Administration)**
Hong Kong SAR, China (Hong Kong Observatory)**
India (India Meteorological Department)**
Japan (Japan Meteorological Agency)**
Korea, Dem. People's Rep. (State Hydrometeorological Administration)
Korea, Rep. (Korea Meteorological Administration)**
Lao PDR (Department of Meteorology)
Macao SAR, China (Meteorological and Geophysical Bureau)
Mongolia (National Agency for Meteorology, Hydrology, and Environmental Monitoring)
Rep. of the Union of Myanmar (Department of Meteorology and Hydrology)
Russian Federation (Russian Federal Service for Hydrometeorology and Environmental Monitoring)**
Thailand (Thai Meteorological Department)
Vietnam (Hydrometeorological Service)

WMO Region V (South West Pacific) countries including the most developed that have regional responsibilities (denoted by ++)

Australia (Bureau of Meteorology)++
Brunei Darussalam (Brunei Darussalam Meteorological Service)
Cook Islands (Cook Islands Meteorological Service)
Fiji (Fiji Meteorological Service)
French Polynesia (Météo France Polynésie Française)
Indonesia (Meteorological, Climatological, and Geophysical Agency)
Kiribati (Kiribati Meteorological Service)
Malaysia (Malaysian Meteorological Department)
Micronesia, Fed. Sts. (FSM Weather Station)
New Caledonia (Météo France Nouvelle Calédonie)
New Zealand (New Zealand Meteorological Service)++
Niue (Niue Meteorological Service)
Papua New Guinea (Papua New Guinea Meteorological Service)
Philippines (Philippine Atmospheric Geophysical and Astronomical Services Administration)

Samoa (Samoa Meteorology Division)
Singapore (Meteorological Service Division)
Solomon Islands (Solomon Islands Meteorological Service)
Timor-Leste (Dirrecão Nacional Meteorologia e Geofisica)
Tonga (Tonga Meteorological Service)
United Kingdom (Met Office)[++]
United States (National Oceanic and Atmospheric
    Administration)[++]
Vanuatu (Vanuatu Meteorological Service)

Weather and Climate Services Progress Model

The following composite criteria are adapted from World Meteorological Organization (WMO) climate services and public weather services and expert opinion, which includes the capacity of the National Meteorological Service (NMS) or National Hydrometeorological Service (NHS) to maintain an observing network and provide forecasts and provide climate services and deliver weather, water, and climate services to users.

Observing and Forecasting Systems

Category 1: Basic Observations and Forecasting
In this category, it is expected that an NMS has the capacity to support a synoptic meteorological network, shares these data on the Global Telecommunication System (GTS), and can provide a minimum weather forecasting capability consisting of at least a one-day forecast based on access to forecasts available on the GTS. NMS has sufficient staff to maintain an observing network but may not operate forecasting on a 24-hour, seven-days-a-week basis. Warnings are limited or not issued.

Category 2: Essential Observations and Forecasting
In this category, it is expected that in addition to capabilities in Category 1, the NMS would also routinely measure the structure of the atmosphere using radiosondes. Automation of observing network is routine. The NMS should also have access to satellite data with the capacity to derive precipitation estimates and the capability to provide flash flood forecasting guidance. The forecasting system should extend from 0 to 3 days based mostly on access to products available on the GTS. If the NMS is responsible for aviation meteorology, it should meet the standards established by WMO and the International Civilian Aviation Organization. If responsible for hydrology, a complete hydrological network of gauges to monitor major rivers is supported. Reliable warnings are routinely

issued. The observing network is sustainable with sufficient budget operating for operations and maintenance.

Category 3: Full Observations and Forecasting

Building on Categories 1 and 2, in this category, observations extend to smaller scales and include ground-based remote sensing techniques, such as radar. Limited area modeling systems are available. Using local data assimilation and numerical models, high-resolution spatially differentiated short-time scale forecasts are produced with emphasis on 0–6 hours for extreme events. Hydrological modeling and flood forecasting on all relevant time scales is routine. The NMS has the capacity to tailor forecasts to specific users. A multihazard warning system exists.

Category 4: Advanced Observations and Forecasting

In addition to the foregoing capabilities, the NMS has an extensive research program and introduces new observational and forecasting technologies and techniques as needed. The NMS has the capacity to support requirements of other NMSs. The observing network is comprehensive, is sufficient to meet main user needs, and may incorporate external observations from other suppliers, for example, agro-meteorological network operated by a Ministry of Agriculture or hydrological network operated by a Ministry of Energy or Water Resources. A high degree of cooperation between government departments and between government and civil society is evident. Forecasts of weather impacts on specific sectors are routine and generally developed with users of these forecasts.

Weather Services Delivery

Category 0: No Service Delivery

NMS has no knowledge of the users or their requirements for products or services. No concept of service delivery, just data or simple products issued. No measures of performance for either accuracy or service delivery are in place. No concept or communication of service delivery principles.

Category 1: Basic Service Delivery

Users are known, but no process for user engagement exists. User requirements for service delivery are not well defined. Services do not respond to changing user needs and new technology. Products are documented with limited descriptive information. Some developing measures are in place to evaluate and monitor performance and outcomes. The verification of accuracy and/or service delivery takes place, but no systematic process exists to use this information to improve the service. The concept of service delivery has been introduced, and an assessment of current status has been undertaken. No formal service delivery training in place, though informal communication of service delivery principles exists.

Category 2: Essential Service Delivery

Users are able to contact NMSs, and their feedback is recorded. There are no formal processes for using the feedback received in development of services. User requirements are defined with limited documentation. Measures of verification and service delivery are in place but are not informed by user requirements. An Action Plan has been created to improve the level of current service delivery and resources have been identified to implement it. A Service Delivery Champion has been identified but does not have appropriate support from all levels of NMHSs to deliver improvements to Service Delivery.

Category 3: Full Service Delivery

NMSs seek input on an ad hoc basis from users to inform development of services. Requirements are defined in documents agreed upon with the customer but are not routinely updated. User feedback is used to inform changes and developments to services. Products and services are consistently documented. Service-Level Agreements are defined. User requirements inform the measures of performance. Findings are used to identify areas for improvement. Subsequent actions are undertaken in an ad hoc manner. An Action Plan is being implemented to improve service delivery, and the outcomes are being monitored. All members of staff are fully aware of the Action Plan and their roles and responsibilities. Formal training is provided. There is an ad hoc process for staff to provide ideas for improvements to service delivery.

Category 4: Advanced Service Delivery

A consistent on-going dialogue is maintained with users in respect of their needs and the services they receive. Requirements are defined in documents agreed upon with the customer, and routinely updated using feedback from users. Users are consulted to inform development of products and services. The service defined in the Service-Level Agreement is agreed upon with the customer based on user consultation. Measures of performance are based on user need, are reported regularly, and are consistently used to inform decisions on improvements. The status of service delivery is reviewed on a regular basis. The Action Plan evolves in response to the outcome of the reviews. There is a culture of providing best possible service delivery. Innovative ideas form a routine input to the Continual Service Improvement process.

Climate Services

Category 1: Basic Climate Services

Functions of NMSs in this category include design, operation, and maintenance of national climate observing systems; data management including quality assurance and quality control; development and maintenance of data archives; climate monitoring; oversight on climate standards; climate diagnostics and climate analysis; climate assessment; dissemination via a variety of media of climate products based on data; participation in regional climate outlooks; and some interaction with users, to meet requests and gather feedback. All NMSs will therefore perform the functions of national climate centers performing the basic climate services. Optimally, staff in Category 1 NMSs should be proficient in climate statistics, homogeneity testing techniques, and quality assurance techniques.

Category 2: Essential Climate Services

In addition to performing all the functions as a national climate center providing basic climate services of Category 1, NMSs should have the capacity to develop and/or provide monthly and longer climate predictions, including seasonal climate outlooks, both statistical and model-based; should be able to conduct or participate in regional and national climate outlook forums; should interact with users in various sectors to identify their requirements for and provide advice on climate information and products; and should get feedback on the usefulness and effectiveness of the information and services provided. Category 2 NMSs would add value from national perspectives on the products received from Regional Climate Centers and in some cases Global Prediction Centers, conduct climate watch programs, and disseminate early warnings. Staff in Category 2 should be proficient in development and interpretation of climate prediction products, and in assisting users in uptake of these products.

Category 3: Full Climate Services

In addition to functions discharged by Category 2 NMSs, the Category 3 NMSs would have the capacity to develop and/or provide specialized climate products to meet the needs of major sectors and should be able to downscale long-term climate projections as well as develop and/or interpret decadal climate prediction (as and when available). These NMSs would meet the requirements for climate information and products to cover all the elements of Climate Risk management, from risk identification, risk assessment, planning and prevention, services for response and recovery from hazards, information relevant to climate variability and change, and information and advice related to adaptation. They would serve to build societal awareness to climate change issues and provide information relevant to policy development and a national action plan. Staff in Category 3 NMSs will require special knowledge in risk and risk management and may have knowledge of financial tools for risk transfer.

Category 4: Advanced Climate Services

In addition to the functions discharged by Category 2 and Category 3 NMSs, the Category 4 NMSs have certain in-house research capacities and would be able to run Global and Regional Climate Models. They would be able to work with sector-based research teams and develop application models (for example, to combine climate and agriculture information and produce food security products), and to develop software and products suites for customized climate products. Staff in Category 4 NMSs will have modeling and statistical expertise in a multidisciplinary context, and will be able to downscale global scale information to regional and national levels. They would also be required to receive and respond to user requirements for new products.

APPENDIX H

Overview of World Bank Activities in East Asia and the Pacific

Figure H.1 Map of World Bank Lending Activities in East Asia and the Pacific

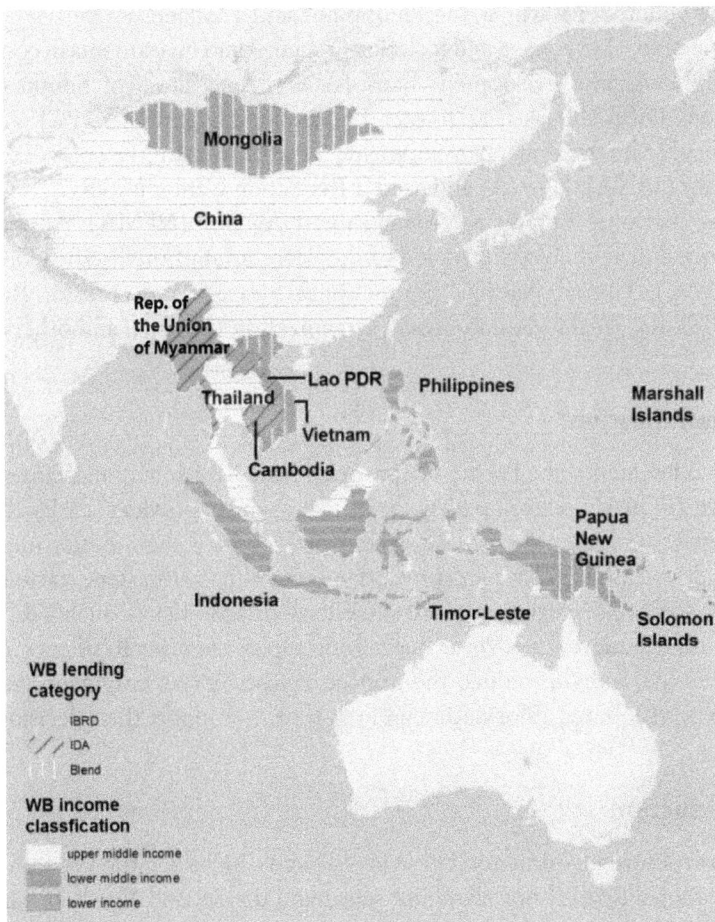

Source: World Bank staff; East Asia and Pacific countries location based on World Bank, ESRI base map, 2012.

Institutional and Capacity Building

- **Capacity building:** Since 2009, more than 1,500 participants from 40 cities in 24 countries across East Asia and the Pacific, and in some sessions, across multiple regions including Africa, the Middle East and North Africa, and South Asia, benefited from 17 learning sessions as part of the Disaster Risk Management (DRM) series open to government officials, policy makers, and practitioners, organized through the Global Development Learning Network together with the World Bank's partners.
- **Knowledge transfer through timely analytical work focusing on key issues:** *Cities and Flooding: A Guide to Integrated Urban Flood Risk Management for the 21ˢᵗ Century* (2012); *Advancing Disaster Risk Financing and Insurance in ASEAN Member States: Framework and Options for Implementation* (2012); *Safer Homes, Stronger Communities: A Handbook for Reconstructing after Natural Disasters* (2010); and *Preparing for Asia's Next Big Earthquake* (2010).
- **Social impact analysis:** This is now a standard part of World Bank–supported post-disaster needs assessments in the region, with experience in the Republic of the Union of Myanmar, the Philippines, and Thailand.
- **Partnerships:** Leveraging and fostering programs and investments in cooperation with the Asian Development Bank (ADB), Association of Southeast Asian Nations (ASEAN), Asian Disaster Preparedness Center (ADPC), Australian Agency for International Development (AusAID), Japan Aerospace Exploration Agency (JAXA), Japan Asian Disaster Reduction Center (ADRC), Republic of Korea National Emergency Management Agency (NEMA), Applied Geo-Science and Technology Division of the Secretariat of the Pacific Community (SOPAC), United Nations International Strategy for Disaster Reduction (UNISDR), World Meteorological Organization (WMO), and others.

Risk Identification

- **Risk assessment:** The Pacific Catastrophe Risk Assessment and Financing Initiative (PCRAFI) developed with SOPAC, which provides 15 Pacific island countries with disaster risk assessment tools to help them better understand, model, and assess their exposure and vulnerability to disasters, started its next phase to develop regional and country-level disaster risk financing solutions.
- **Risk information:** The World Bank–wide Open Data for Resilience Initiative (OpenDRI) aims to reduce the impact of disasters by empowering decision makers with better information and the tools to support their decisions.

Risk Reduction

- **Infrastructure investments:** These investments, including the Western Indonesia National Roads Improvement Project and the Second Northern Mountains Poverty Reduction Project in Vietnam, have components with built-in resilience measures.

- **Urban resilience:** The Building Urban Resilience Program, funded by the Aus-AID East Asia Infrastructure Growth Fund, supports city-level investment decisions by identifying the key drivers of risk to disasters and climate change, and creating a set of web-based open-source risk assessment tools used in decision making by city-level institutions, private investors, communities, and planners of infrastructure services.
- **Linking DRM and climate change adaptation:** The Bank does this through investments, such as the Adaptation Projects in Kiribati and the Solomon Islands.

Emergency Preparedness

- **Early warning systems:** With WMO and UNISDR, the Bank is supporting an assessment of hydromet services in Cambodia, Indonesia, the Lao People's Democratic Republic, the Philippines, and Vietnam to better collect, process, and disseminate hydromet hazard data.
- **Risk management tools:** The InaSAFE platform and suite of analytical tools, prepared by the GFDRR Labs and the Australia-Indonesia Facility for Disaster Reduction in partnership with the Indonesian National Disaster Management Agency helps disaster risk managers with contingency planning and investment decisions.

Financial Protection

- **Budget support:** In the Philippines, a DRM development policy loan with a Catastrophe Deferred Drawdown Option worth US$500 million offers immediate liquidity to mitigate the government's fiscal exposure to a disaster.
- **Quick disbursing contingent financing components:** These are now routinely embedded in investment projects in East Asia and the Pacific, such as in Lao PDR, that provide rapid investment funding in the event of a major natural disaster.
- **Insurance:** In Mongolia, an Index-Based Livestock Insurance Program has supported the establishment and scaling up of a new approach to insurance for weather-related livestock losses, which includes an International Development Association-financed contingent debt facility for rapid drawdown.
- **Risk modeling:** In Indonesia, the Bank is working with the Ministry of Finance's Fiscal Risk Office to better model the contingent risks of disasters.

Sustainable Recovery and Reconstruction

- **Post-disaster assessment:** Between 2008 and 2012, in partnership with the European Union and the United Nations, the Bank has supported the governments of Cambodia, Indonesia, Lao PDR, the Philippines, Samoa, and Thailand and ASEAN in the Republic of the Union of Myanmar to conduct post-disaster needs analyses to support rapid, sustainable recovery and reconstruction planning, and to mobilize international funding.

- **Reconstruction financing:** The Bank partnered with the government of Indonesia as part of the Aceh reconstruction by managing the US$685 million multidonor trust fund. Current lending projects include the China Wenchuan Earthquake Recovery Project, Samoa Post-Tsunami Reconstruction Project, Tonga Post-Tsunami Reconstruction Project, Typhoon Ketsana Reconstruction (Cambodia), and the Vietnam Natural Disaster Management Project (Additional Financing).

- **Community-driven resilient reconstruction:** The Bank rapidly mobilized US$3.5 million of grants through the Java Reconstruction Fund to finance the reconstruction and rehabilitation of housing and community infrastructure after the eruption of Mount Merapi through the existing community-driven development program (REKOMPAK). In Vietnam the model of community-based DRM piloted under the First National DRM Project has been scaled up nationally. The Tonga Post-Tsunami Reconstruction Project supports the construction of cyclone-resistant housing with equipment for hazard and risk assessment, institutional strengthening of the planning and global information services units, and preparation of community DRM plans. The Samoa Post-Tsunami Reconstruction Project supports community coastal infrastructure management plans.

Glossary of Key Terms

All terminology is taken from the IPCC glossary (IPCC 2012), apart from "risk assessment," which is taken from the 2009 UNISDR glossary, and "fragility," which follows the World Bank definition.

Adaptation: In human systems, the process of adjustment to actual or expected climate and its effects, in order to moderate harm or exploit beneficial opportunities. In natural systems, the process of adjustment to actual climate and its effects; human intervention may facilitate adjustment to expected climate.

Capacity: The combination of all the strengths, attributes, and resources available to an individual, community, society, or organization that can be used to achieve established goals.

Climate change: A change in the state of the climate that can be identified (for example, by using statistical tests) by changes in the mean and/or the variability of its properties and that persists for an extended period, typically decades or longer. Climate change may be due to natural internal processes or external forces, or to persistent anthropogenic changes in the composition of the atmosphere or in land-use.

Climate extreme (extreme weather or climate event): The occurrence of a value of a weather or climate variable above (or below) a threshold value near the upper (or lower) ends of the range of observed values of the variable. For simplicity, both extreme weather events and extreme climate events are referred to collectively as "climate extremes."

Disaster: Severe alterations in the normal functioning of a community or a society due to hazardous physical events interacting with vulnerable social conditions, leading to widespread adverse human, material, economic, or environmental effects that require immediate emergency response to satisfy critical human needs and that may require external support for recovery.

Disaster risk: The likelihood over a specified time period of severe alterations in the normal functioning of a community or a society due to hazardous physical events interacting with vulnerable social conditions, leading to widespread adverse human, material, economic, or environmental effects that require immediate emergency response to satisfy critical human needs and that may require external support for recovery.

Disaster risk management: Processes for designing, implementing, and evaluating strategies, policies, and measures to improve the understanding of disaster risk, foster disaster risk reduction and transfer, and promote continuous improvement in disaster preparedness, response, and recovery practices, with the explicit purpose of increasing human security, well-being, quality of life, and sustainable development.

Disaster risk reduction: Denotes both a policy goal or objective, and the strategic and instrumental measures employed for anticipating future disaster risk; reducing existing exposure, hazard, or vulnerability; and improving resilience.

Early warning system: The set of capacities needed to generate and disseminate timely and meaningful warning information to enable individuals, communities, and organizations threatened by a hazard to prepare and to act appropriately and in sufficient time to reduce the possibility of harm or loss.

Fragility: A fragile situation is defined as having either (1) a composite World Bank, African Development Bank, and Asian Development Bank Country Policy and Institutional Assessment rating of 3.2 or less or (2) the presence of the United Nations and/or regional peace-keeping or peace-building mission (for example, African Union, European Union, NATO), with the exclusion of border-monitoring operations, during the last three years.

Governance: This can be understood as the structures of common governance arrangements and processes of steering and coordination—including markets, hierarchies, networks, and communities. Formal and informal governance structures also determine vulnerability because they influence power relations and risk perceptions, and constitute the context in which vulnerability, risk reduction, and adaptation are managed. Countries with institutional and governance fragilities often lack the capacity to identify and reduce risks and to deal with emergencies and disasters effectively.

Hazard: The potential occurrence of a natural or human-induced physical event that may cause loss of life, injury, or other health impacts, as well as damage and loss to property, infrastructure, livelihoods, service provision, and environmental resources.

Mitigation (of disaster risk and disaster): The lessening of the potential adverse impacts of physical hazards (including those that are human-induced) through actions that reduce hazard, exposure, and vulnerability.

Natural hazard: The potential occurrence of a natural physical event that may cause loss of life, injury, or other health impacts, as well as damage and loss to property, infrastructure, livelihoods, service provision, and environmental resources.

Preparedness: The knowledge and capacities developed by governments, professional response and recovery organizations, communities, and individuals to effectively anticipate, respond to, and recover from the impacts of likely, imminent, or current hazard events or conditions.

Prevention: Disaster prevention expresses the concept and intention to completely avoid potential adverse impacts through action taken in advance. Examples include dams or embankments that eliminate flood risks, land-use regulations that do not permit any settlement in high-risk zones, and seismic engineering designs that ensure the survival and function of a critical building in any likely earthquake. Very often the complete avoidance of losses is not feasible, and the task transforms to that of mitigation. Partly for this reason, the terms prevention and mitigation are sometimes used interchangeably in casual use.

Recovery (days to months): The situation returns to a relatively normal state (but not normality). Throughout reconstruction (months to years), the region is slowly, fully returned to normality.

Resilience: The ability of a system and its component parts to anticipate, absorb, accommodate, or recover from the effects of a hazardous event in a timely and efficient manner, including through ensuring the preservation, restoration, or improvement of its essential basic structures and functions.

Response (0–10 days) **and Relief** (0–25 days): The situation is stabilized, including rescue, immediate medical aid, food and emergency shelter provisions, dead are cared for, and dangerous structures and situations are identified and isolated or controlled.

Risk assessment: A methodology to determine the nature and extent of risk by analyzing potential hazards and evaluating existing conditions of vulnerability that together could potentially harm exposed people, property, services, livelihoods, and the environment on which they depend.

Risk governance: Describes a process of exchanging, integrating, and sharing knowledge and information, which engages a wide range of stakeholder groups, such as scientists, policy makers, private firms, nongovernmental organizations, media, educators, and the public. The risk governance framework offers a systematic way to help situate such judgments about disaster management, risk reduction, and risk transfer. Risk governance consists of four phases—pre-assessment, appraisal, characterization/evaluation, and management—in an open, cyclical, iterative, and interlinked process. Risk communication accompanies all four phases. This process is consistent with those in the UNISDR Hyogo Framework for Action. Risk governance uses concepts from probabilistic risk analysis to help judge appropriate allocations in level of effort and over time and among risk reduction, risk transfer, and disaster management actions.

Risk transfer: The process of formally or informally shifting the financial consequences of particular risks from one party to another whereby a household, community, enterprise, or state authority will obtain resources from the other party after a disaster occurs, in exchange for ongoing or compensatory social or financial benefits provided to that other party.

Reference

IPCC. 2012. *Managing the Risks of Extreme Events and Disasters to Advance Climate Change Adaptation.* Special Report of Working Groups I and II of the Intergovernmental Panel on Climate Change. Edited by C. B. Field, V. Barros, T. F. Stocker, D. Qin, D. J. Dokken, K. L. Ebi, M. D. Mastrandrea, K. J. Mach, G.-K. Plattner, S. K. Allen, M. Tignor, and P. M. Midgley. Cambridge, UK: Cambridge University Press.

Environmental Benefits Statement

The World Bank is committed to reducing its environmental footprint. In support of this commitment, the Office of the Publisher leverages electronic publishing options and print-on-demand technology, which is located in regional hubs worldwide. Together, these initiatives enable print runs to be lowered and shipping distances decreased, resulting in reduced paper consumption, chemical use, greenhouse gas emissions, and waste.

The Office of the Publisher follows the recommended standards for paper use set by the Green Press Initiative. Whenever possible, books are printed on 50% to 100% postconsumer recycled paper, and at least 50% of the fiber in our book paper is either unbleached or bleached using Totally Chlorine Free (TCF), Processed Chlorine Free (PCF), or Enhanced Elemental Chlorine Free (EECF) processes.

More information about the Bank's environmental philosophy can be found at http://crinfo.worldbank.org/crinfo/environmental_responsibility/index.html.

green
press
INITIATIVE

www.ingramcontent.com/pod-product-compliance
Lightning Source LLC
Chambersburg PA
CBHW080611270326
41928CB00016B/3013